Painting Guide for the Boeing Stratofortress Motherships

Brian Lockett

LockettBooks
www.Air-and-Space.com

Acknowledgements:

Boeing Archives:
> Mike Lombardi

Air Force Flight Test Center History Office (AFFTC/HO):
> Ray Puffer, Dr. James Young

NASA Dryden Flight Research Center:
> Dill Hunley, Peter Merlin, Tony Landis

Bob Dorr, Greg Spahr, Terry Panopalis

Dedicated to my father,

Richard Lockett,

who inspired my interest in all things aeronautical.

About the author:

Brian Lockett earned his Private Pilot certificate as a teenager. He has been a member of the American Aviation Historical Society since 1973, and the Journal of the A.A.H.S. has published seven of his aviation history articles. He graduated from the University of California at Santa Barbara with a Bachelor of Arts degree in Geography. Brian worked for nearly a decade at the Santa Barbara County Air Pollution Control District. He is the creator of the Goleta Air & Space Museum web site (www.Air-and-Space.com). He is the author of *Flying Aircraft Carriers of the USAF: Project FICON* and *Balls Eight: History of the Boeing NB-52B Stratofortress Mothership*. His username is GoletaBrian on the Gigapan, Panoramio, Weather, and YouTube web sites. He publishes calendars, DVDs, and books as LockettBooks at Lulu.com.

Also by Brian Lockett:

> *Flying Aircraft Carriers of the USAF: Project FICON*
> *Flying Aircraft Carriers of the USAF: Wing Tip Coupling*
> *Balls Eight: History of the Boeing NB-52B Stratofortress Mothership*

Front cover: Side views of the NB-52B 52-0008 as it appeared in May 1960, May 1962, and November 2004.

Rear cover: Side views of the NB-52A 52-0003 as it appeared in April 1959, April 1961, and May 1969.

LockettBooks
www.Air-and-Space.com

©2009, Brian Lockett
First Edition

ISBN 978-0-578-03110-1

Contents

Introduction 1
  NB-52 Models and Decals 1

NB-52A 52-0003 2
  @ 1953 2
  @ September 1955 6
  @ December 1958 9
  @ April 1959 13
  @ May 1959 17
  @ August 1959 19
  @ April 1960 21
  @ August 1960 23
  @ October 1960 26
  @ April 1961 29
  @ May 1962 32
  @ May 1963 35
  @ May 1965 38
  @ May 1969 41

NB-52B 52-0008 45
  @ May 1960 45
  @ September 1960 49
  @ October 1960 52
  @ May 1961 55
  @ November 1961 58
  @ May 1962 61
  @ May 1963 64
  @ 1964 67
  @ October 1965 69
  @ May 1966 72
  @ 1971 74
  @ May 1972 75
  @December 1974 77
  @December 1977 79
  @ 1979 81
  @ June 1995 82
  @ April 2001 85
  @ November 2004 88

Introduction

This document tracks the changes to the appearance of the NB-52A and NB-52B during their service lives. Two Boeing B-52 Stratofortresses were modified to carry and launch the North American X-15 rocket planes, NB-52A 52-0003 and NB-52B 52-0008. The NB-52A first launched the X-15 on a glide flight on June 8, 1959. The NB-52B first launched an X-15 flight on January 23, 1960. The two NB-52s went on to launch the modified X-15A-2, Northrop HL-10, Northrop M2-F2, and Martin-Marietta X-24A.

The NB-52A retired in October 1969, but the NB-52B soldiered on until November 2004. It launched the Northrop M2-F2/F3 and Martin Marietta X-24B lifting bodies. It simulated the steep, power off approach to landing used by the Space Shuttles. It assisted in the collection of data about wake turbulence from large aircraft. It served as an air-to-air gunnery target. It launched 3/8-scale F-15 Remotely Piloted Research Vehicles (RPRV) and Spin Research Vehicles (SRV). It launched a Ryan Firebee II drone and the Ryan Firebee based Drones for Aeroelastic Structures Testing (DAST). It launched the Highly Maneuverable Aircraft Technology (HiMAT) RPRVs. It dropped the 48,000-pound Space Shuttle Reusable Booster Drop Test Vehicle (SRB/DTV). It released a simulated F-111 crew module from its bomb bay to evaluate new parachute recovery systems. It was the first airplane to launch a satellite into orbit on the Orbital Sciences Pegasus booster. It tested the drag chute used to decelerate space shuttle orbiters. It tested pollution reducing fuel additives with a pair of jet engines mounted under its bomb bay. It launched the X-38 Space Station Crew Return Vehicles. It launched the X-43A Hyper-X Supersonic Combustion Ramjets.

The Stratofortress motherships are popular subjects for modelers. Their special missions capture the imagination. The liberal application of DayGlo orange, DayGlo red, and yellow makes them a couple of the most colorful B-52s.

The appearance of the NB-52s changed many times over the years. Initially, both of them sported bare aluminum with large areas of fluorescent DayGlo red and DayGlo orange. The distribution of the DayGlo was revised occasionally. Eventually both airframes were painted silver and the DayGlo disappeared. The last major change occurred when the NB-52B acquired a yellow NASA tail band in 1976. These changes are illustrated in the following pages. There are fourten sets of illustrations for the NB-52A and eighteen sets of illustrations for the NB-52B.

The frequent alteration in the appearnce of the NB-52s has made it diffficult for modelers to accurately reproduce them in scale. It is hoped that this document will help modelers to reproduce the correct appearance of either Stratofortress for any particular mission.

NB-52 Models and Decals

The Stratofortress motherships are such a popular subject for modeling that a couple of injection molded plastic kits of them have been released. Revell made a small scale kit of the NB-52 with an X-15 in the early 1960s. Monogram released a 1/72-scale kit of the NB-52B with the X-15A-2 in the mid-1980s. Dragon issued a 1/200-scale kit of the NB-52B with the North American X-15A-2 and the Orbitl Sciences Pegasus booster. All three kits are long out of production.

Cutting Edge Decals released decal sheets for both the NB-52A and the NB-52B in 1/72 and 1/144 scale. They include multiple sets of mission marks, reflecting the occasional rearranging of the marks to make room for new ones. The Cutting Edge decals served as reference for the placement of some of the mission marks depicted in these ilustrations. Cutting Edge also released a resin conversion kit for the Crown/Minicraft 1/144-scale B-52H model.

Corgi released a 1/144-scale die cast model of the NB-52B with an X-15 modeled on its appearance in May 1960 when it carried the maximum amount of DayGlo. Model Power released a 1/300-scale die cast model of the NB-52A modeled on its appearance in late 1959 or 1960.

These illustrations are based on line drawings provided by the Boeing Aerospace Company. The side views are presented at 1/200-scale and the top and bottom views are presented at 1/250-scale.

Because it is not possible to determine the precise date for each alteration in the appearance of these Stratofortresses, the dates given below represent the date of the earliest available photographs for each. The changes described may actually have occurred at some earlier time.

NB-52A 52-0003

@ 1953

Initially, B-52A 52-003 was used as a service test vehicle. Like the other two B-52As, it was built without wing tip tanks. It was equipped with the standard four 50-caliber machine gun-turret that was used on most other B-52s.

52-003 carried the usual set of United States Air Force markings. The words "UNITED STATES AIR FORCE" appeared on either side of the forward fuselage in 12-inch tall letters. Black anti-glare panels were painted under the windows of the cockpit and behind the windows of the tail turret station. The top of each wing tip was black.

The nose radome was reddish brown. The chin and belly radomes were black. The Air Force registration appeared on the tail as 2003.

The primary difference from standard B-52 markings was a photo-theodelite target on the upper fuselage just behind the wings. This took the form of a black rectangle with a white bar draped over the top of the fuselage with a horizontal white bar on either side of the fuselage. Viewed from the side, it appeared as a white cross on a black background.

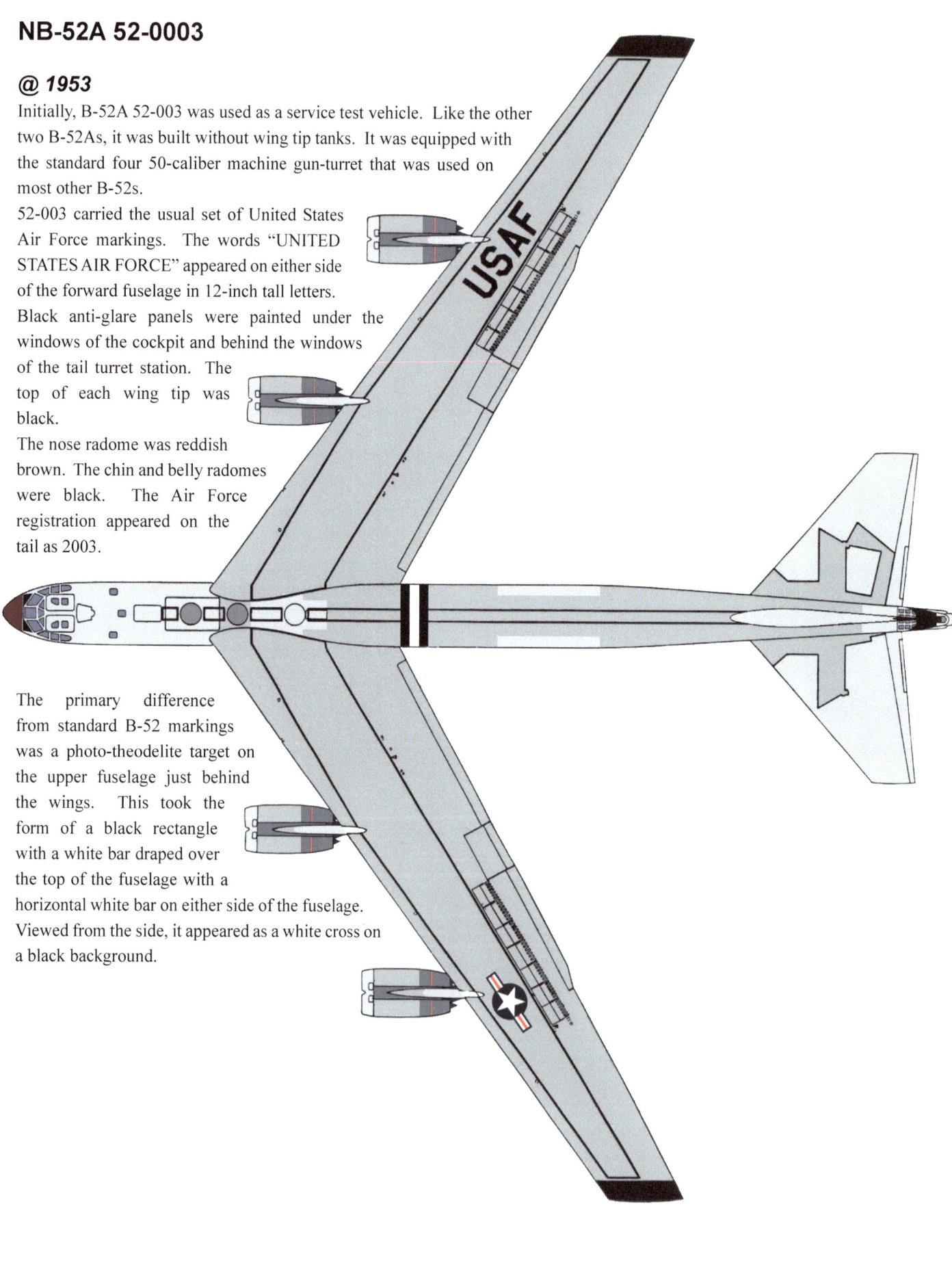

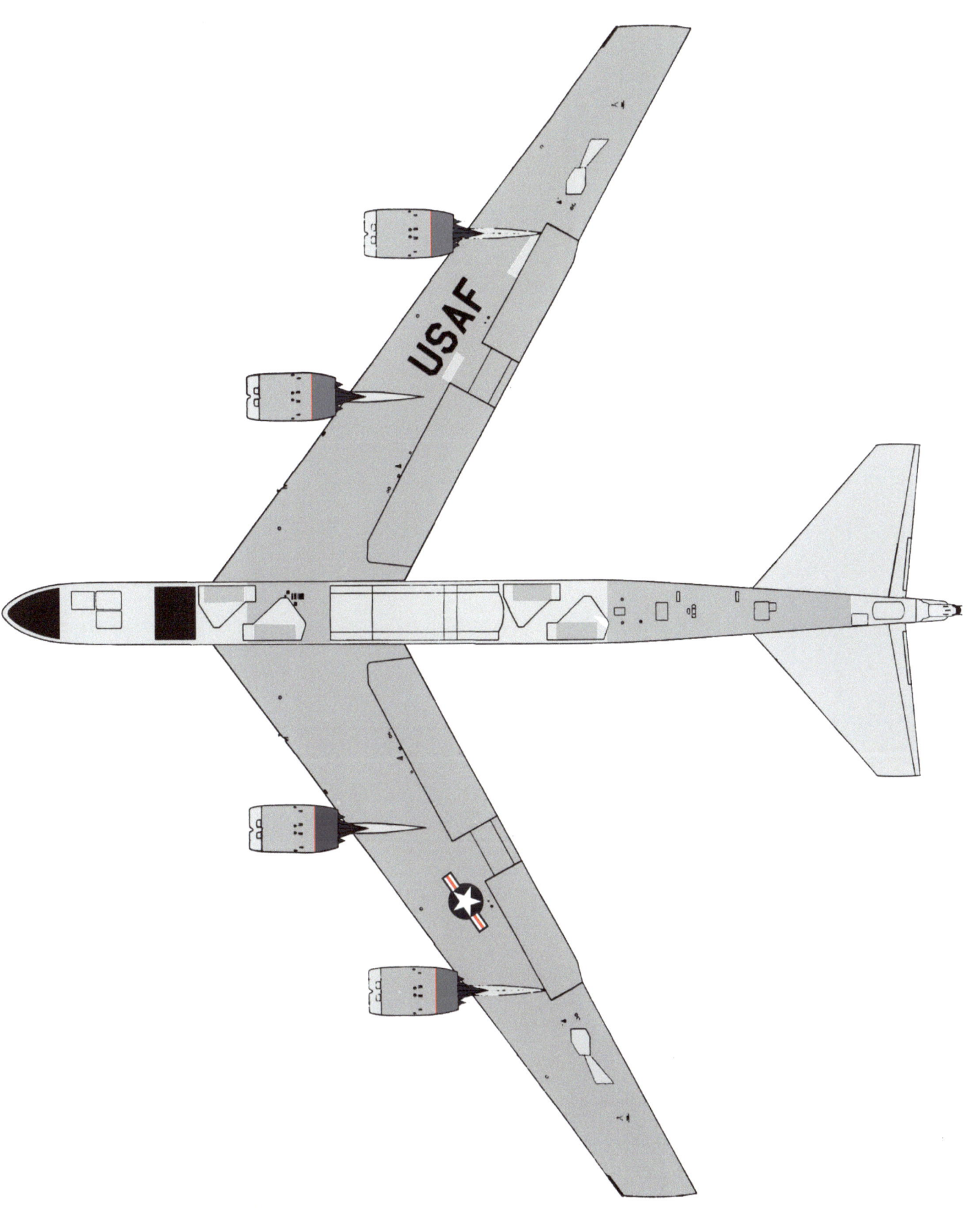

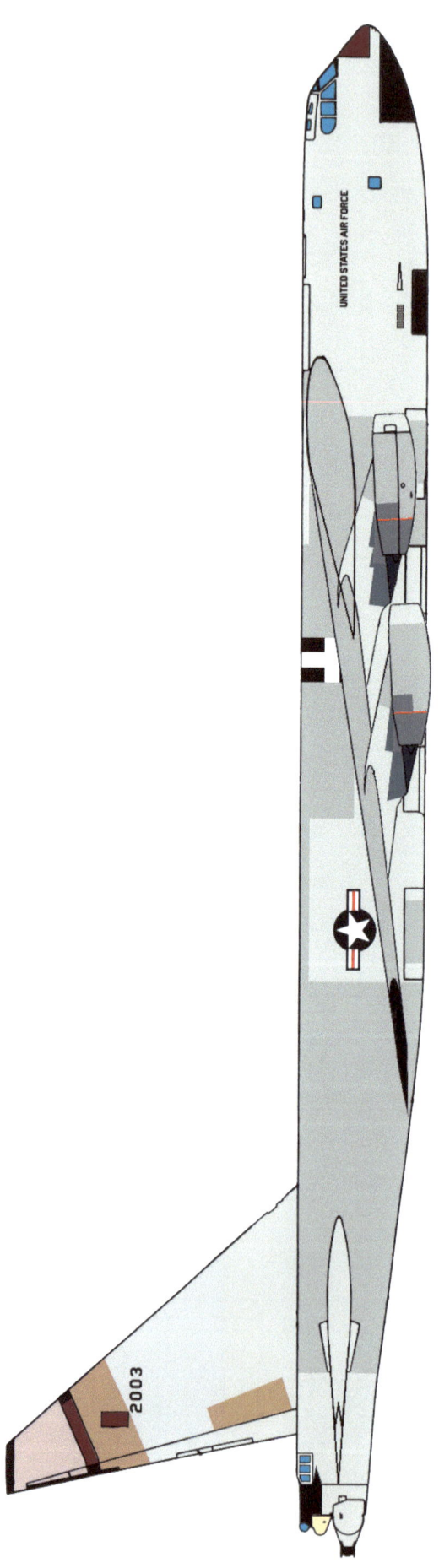
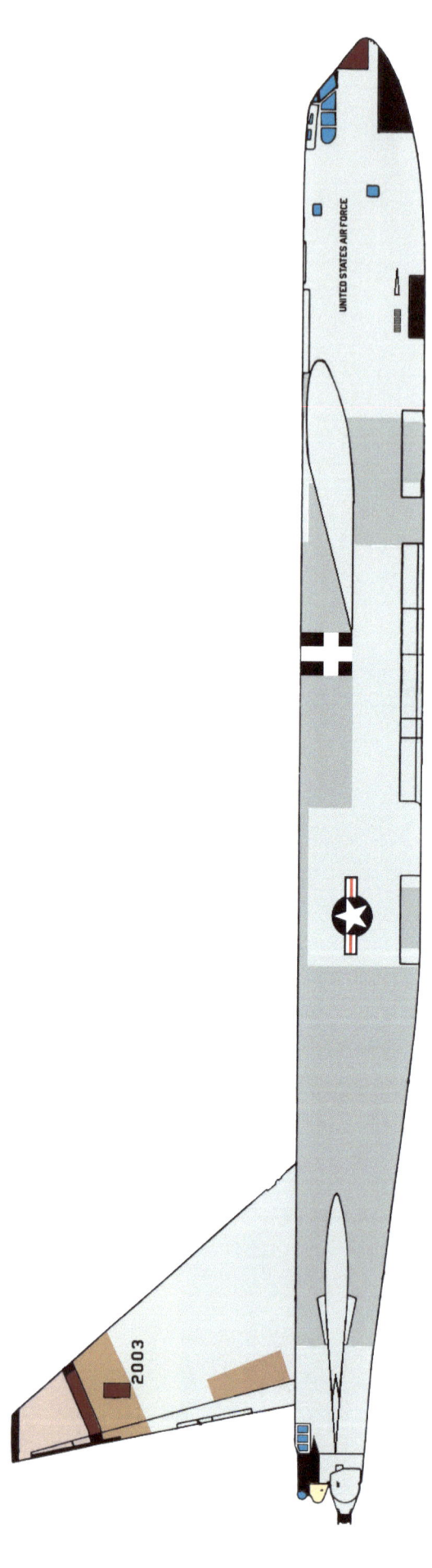

The skin of the B-52 is composed partly of shiny, unpainted aluminum and partly of another alloy that is covered in s duller surface coating. The differing surface material is represented by two shades of gray in these illustrations.

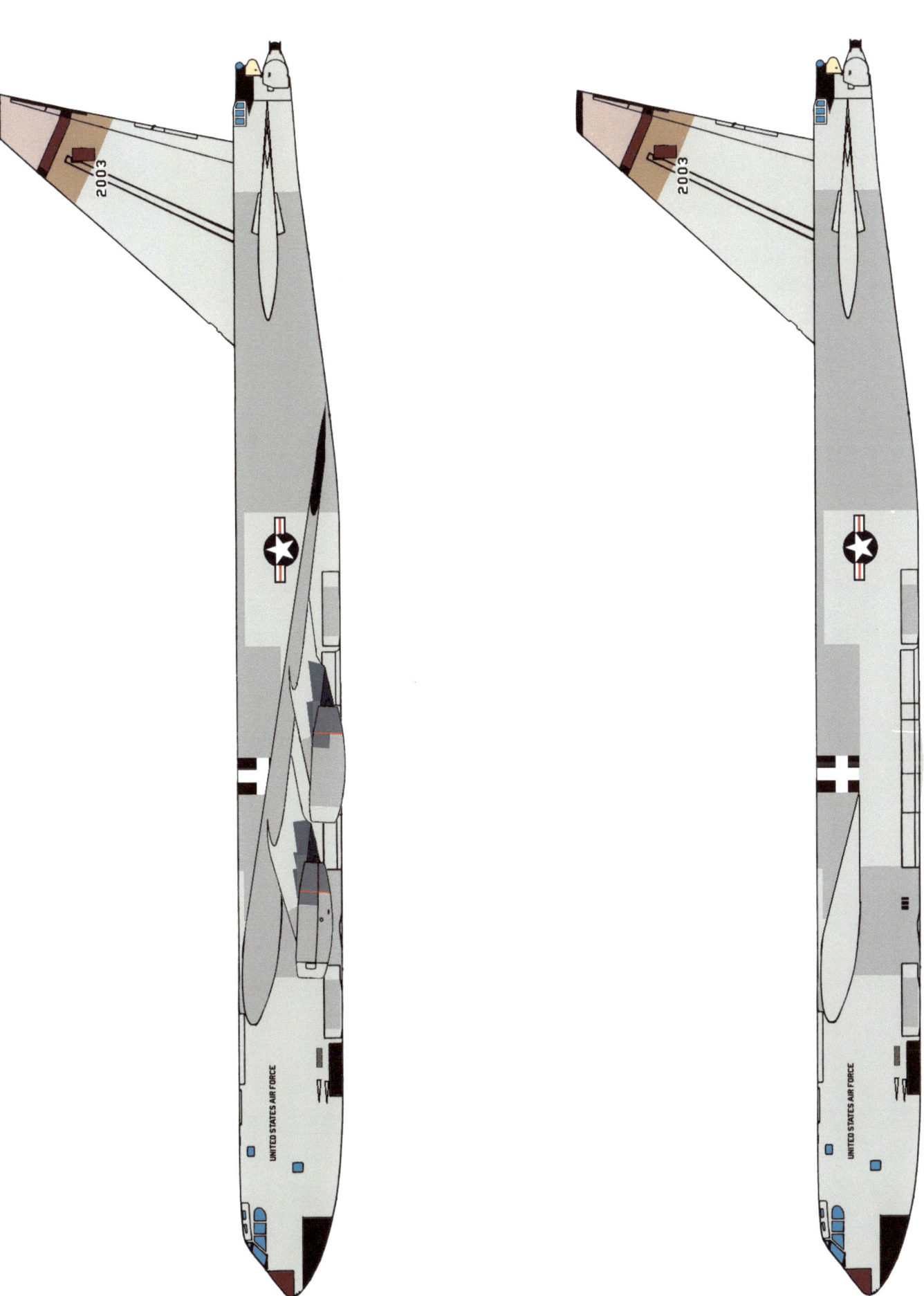

@ September 1955

52-003 was assigned to the Air Force Flight Test Center in 1955. The designation of the aircraft was changed to JB-52A to reflect its modification from service standard.

Wing tip tanks of the type used on the B-52B were added.

A rectangular black area was painted across the upper fuselage in the area of the ECM Operator's station.

The Air Force Flight Test Center shield was applied below and behind the cockpit windows and the last three digits of the tail number were repeated next to the badge.

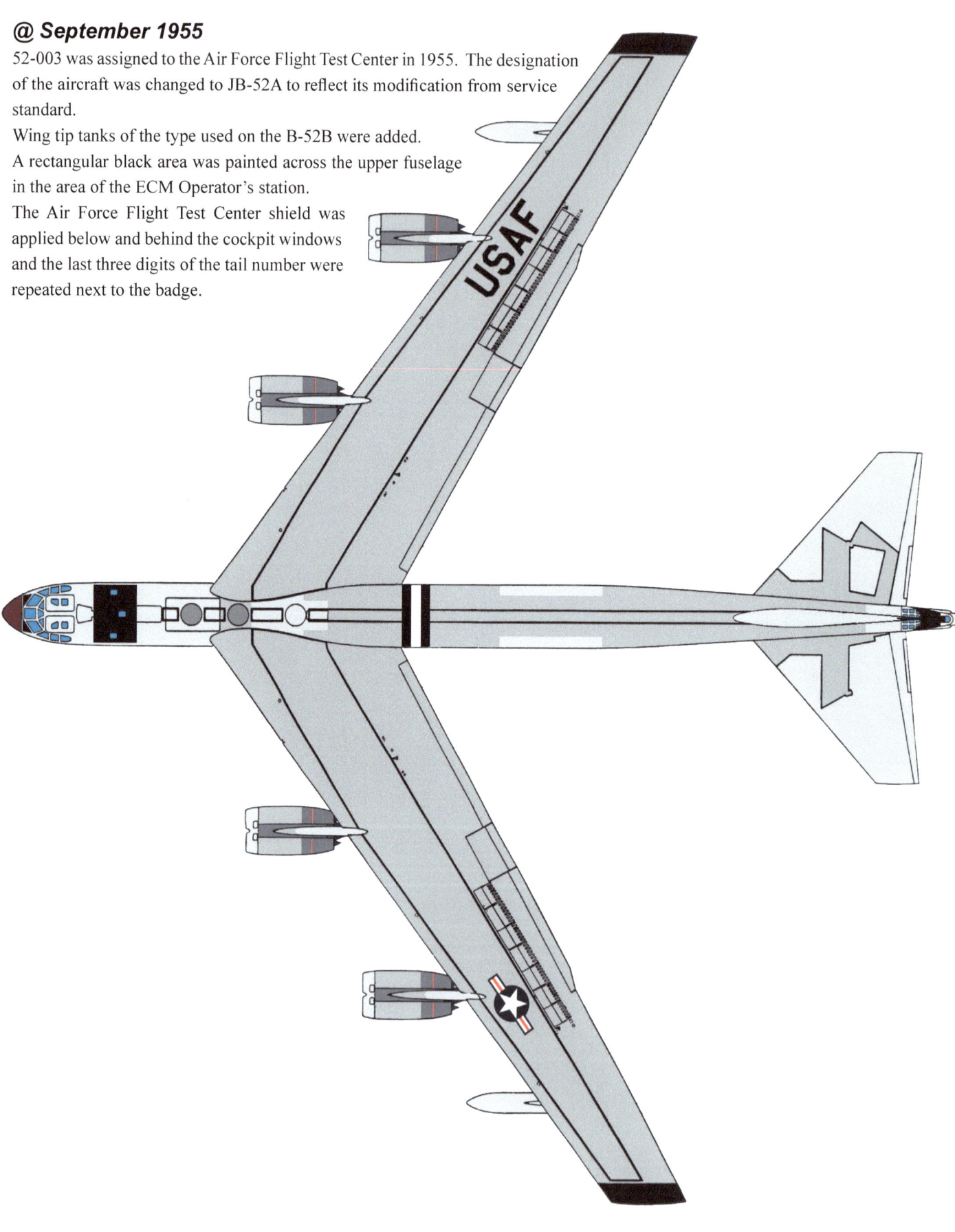

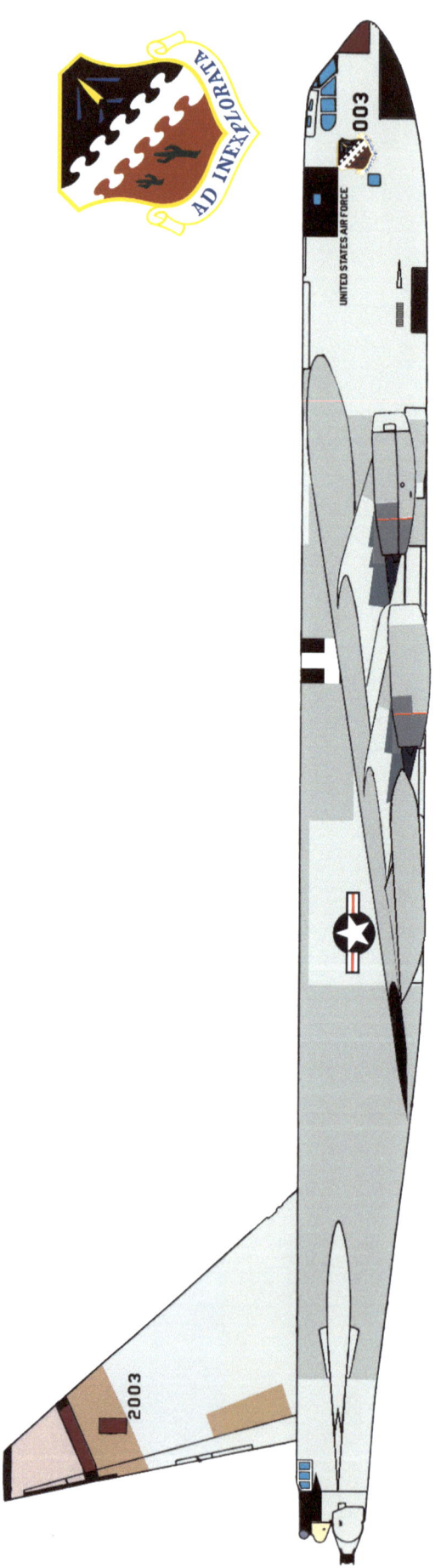
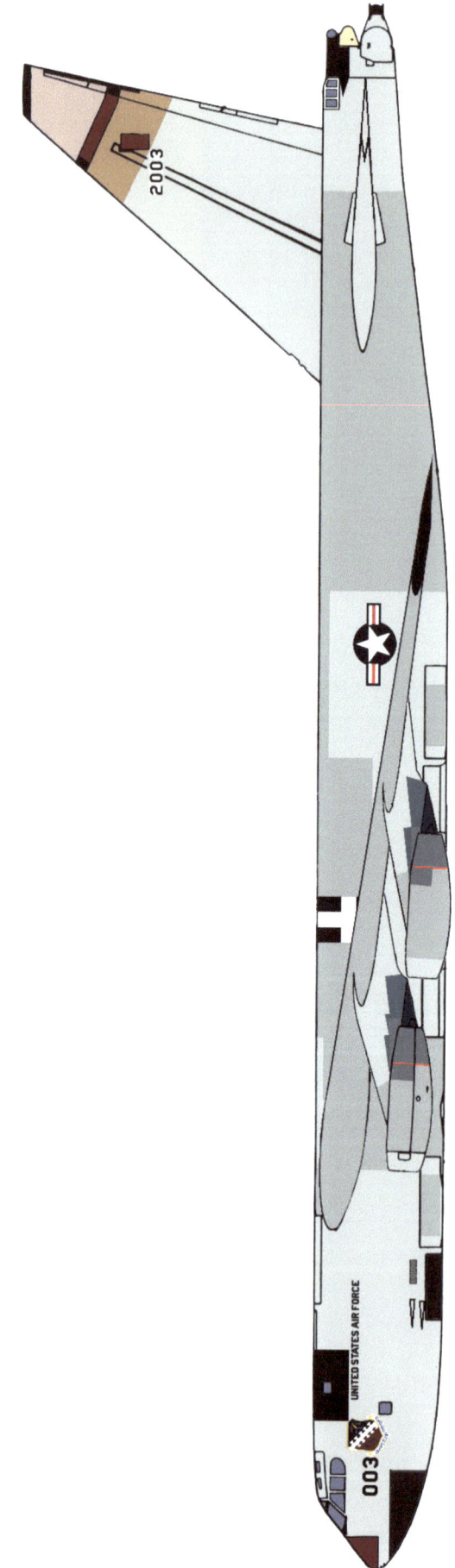

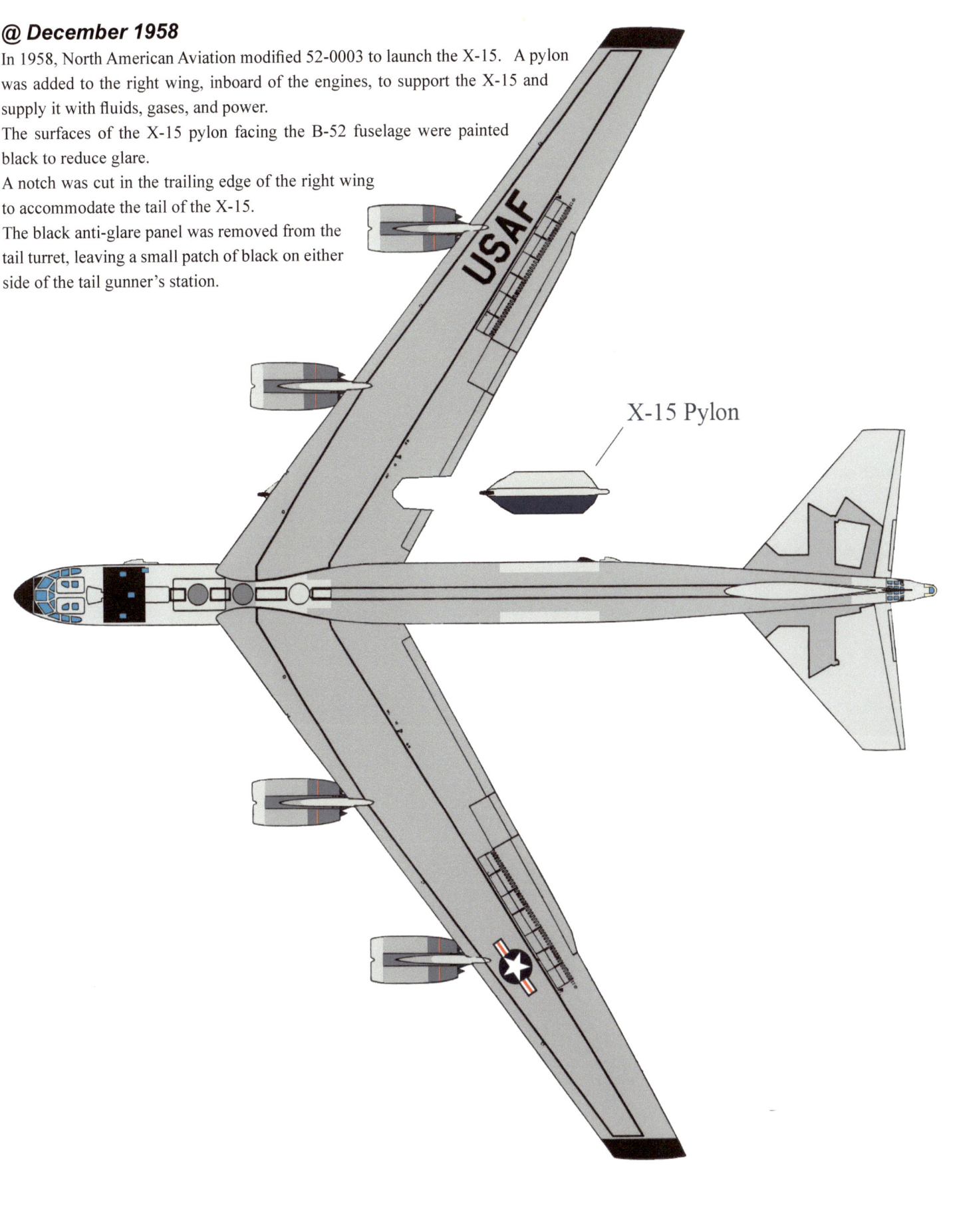

@ December 1958

In 1958, North American Aviation modified 52-0003 to launch the X-15. A pylon was added to the right wing, inboard of the engines, to support the X-15 and supply it with fluids, gases, and power.

The surfaces of the X-15 pylon facing the B-52 fuselage were painted black to reduce glare.

A notch was cut in the trailing edge of the right wing to accommodate the tail of the X-15.

The black anti-glare panel was removed from the tail turret, leaving a small patch of black on either side of the tail gunner's station.

X-15 Pylon

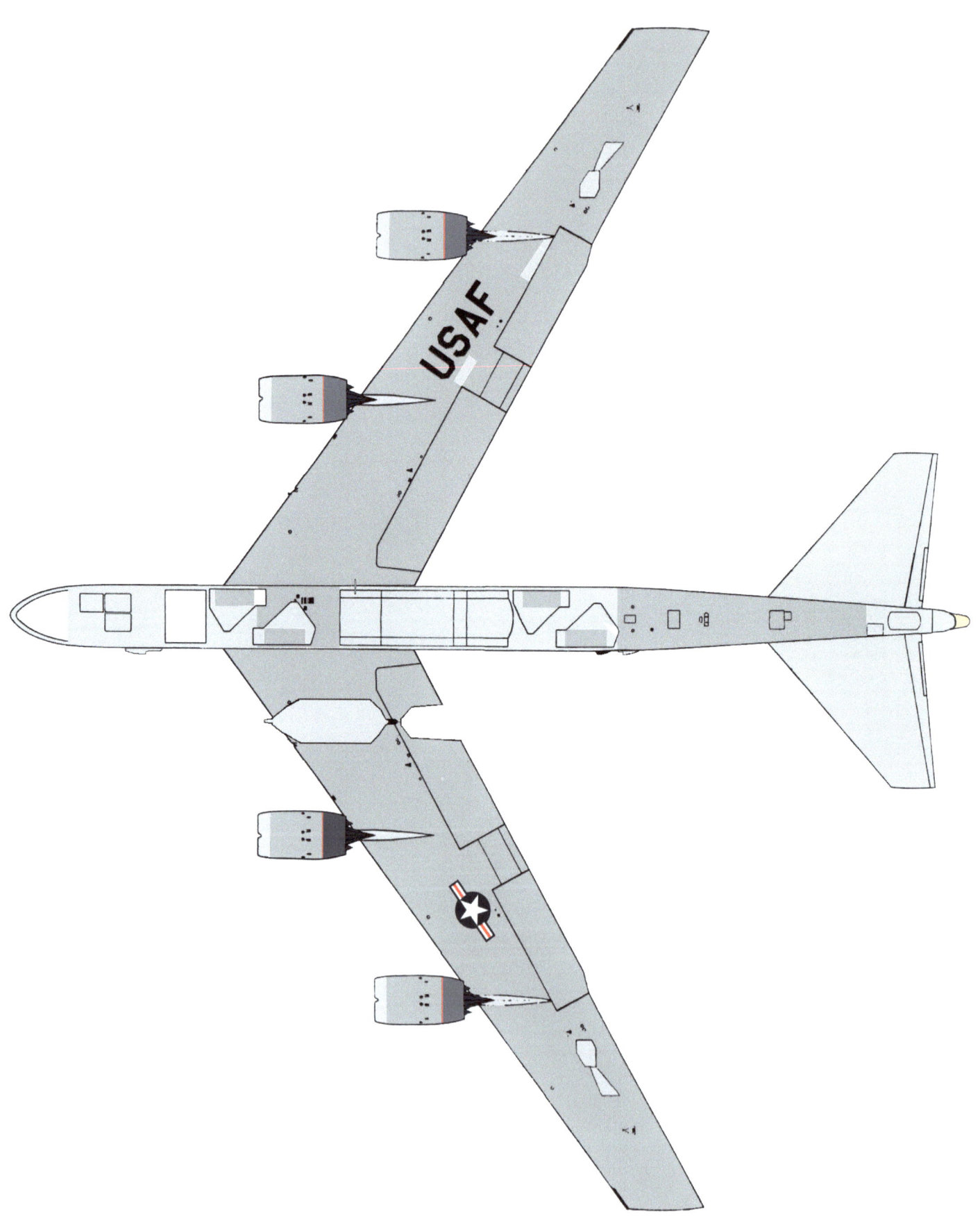

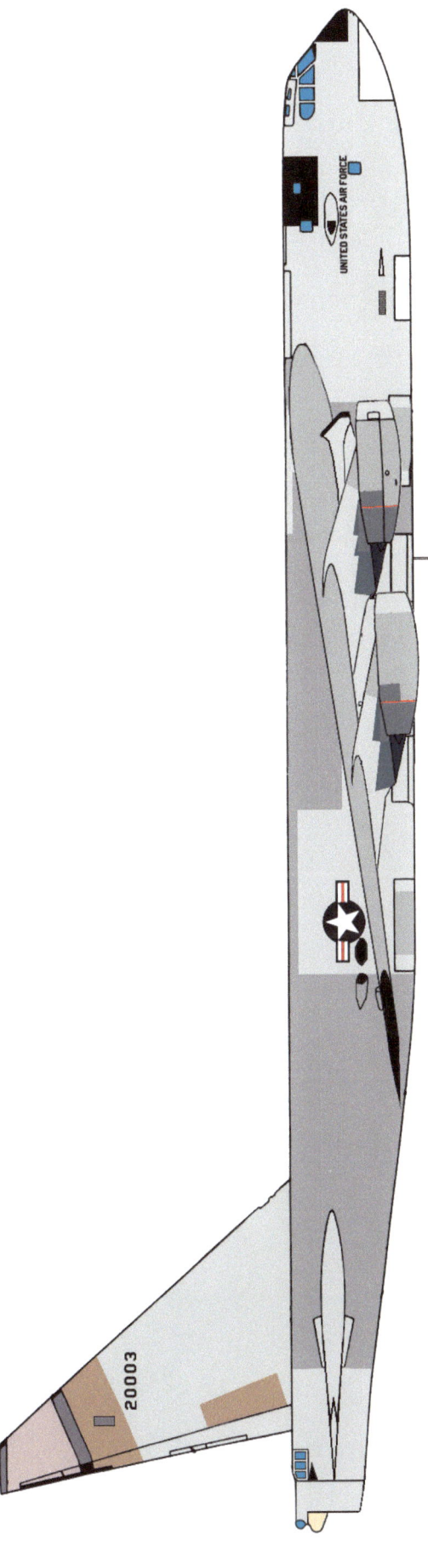

The tail turret was removed, and the space was faired over.
Housings for video and motion picture film cameras were added to the right side of the fuselage.
An additional window was installed at the launch panel operator's station on the right side of the fuselage.
Three floodlights were flush-mounted in the right side of the fuselage. A forward facing floodlights was mounted in a housing directly below the aft video camera.
Ten vents were added to either side of the bomb bay. A liquid oxygen jettison pipe extended from the forward end of the bomb bay on the left side of the fuselage. Two liquid oxygen overflow vents were installed on the left side of the bomb bay.
The photo theodelite target was removed from the upper fuselage.
The AFFTC badge and the "003" were removed from the forward fuselage.
The words "UNITED STATES AIR FORCE" on the forward fuselage were moved down to a position below the video camera housing.
The nose radome was replaced with a black radome. The chin and belly radomes were replaced with white radomes.
An additional zero was added to the tail number, so that it read 20003.
A black, striped photo-resolution target on a white background was painted on the right side of the fuselage below the trailing edge of the wing. A smaller black rectangle appeared directly below the white rectangle.
A small, red square surrounded the base of the liquid oxygen vents and jettison pipe on the left side of the bomb bay.

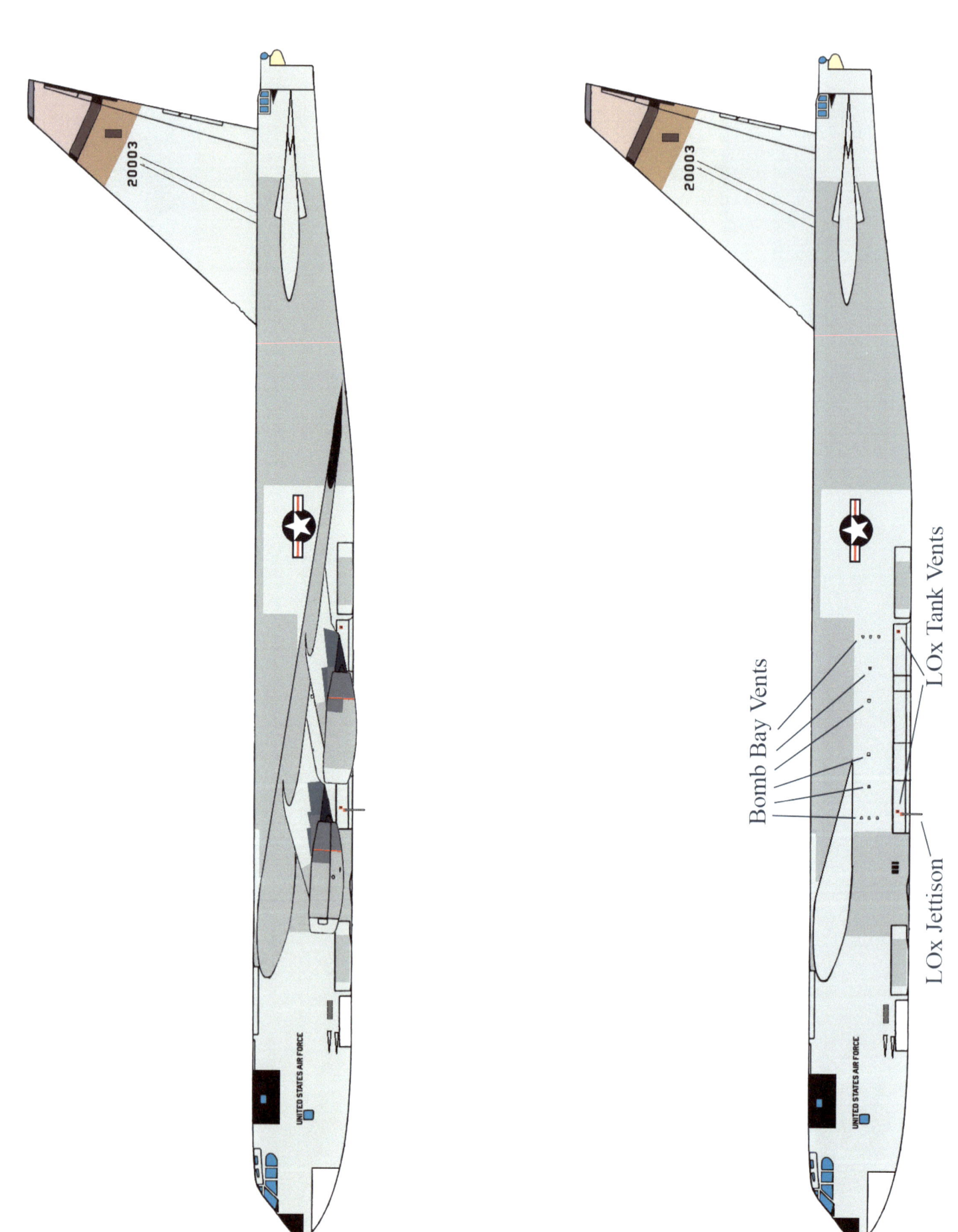

@ April 1959

Following the structural testing of the X-15 mating devices in late 1958, large areas of the nose, tail, wingtips, and engine inlets of the NB-52A were painted Day-Glo orange . Day-Glo orange was used extensively on the aircraft operated from Edwards AFB in that era because it contrasted sharply with the dark blue desert sky and made tracking airplanes in flight easier. Also, orange wreckage is much easier to spot in the desert than aluminum, which tends to mirror the color of its surroundings.

The Day-Glo around the cockpit raked down along the sides of the fuselage, all the way around the bottom. The trailing edge of the orange area ran from the forward upper corner of the lower deck window to the outboard, rear corner of the aft window in each of the pilots' escape hatches.

The windshield frames on either side of the center panel of the windshield were painted matte black.

Each wingtip was painted black on the top and white on the bottom. A broad area of Day-Glo was painted inboard of the wingtip stripes on both the top and bottom of the wings. The flaps of just the right wing were painted white.

The leading edge of the right wing was painted black for several feet on either side of the X-15 pylon. On the upper surface of the wing the inboard edge of the anti-glare panel was clipped perpendicular to leading edge of the wing. The outboard edge was clipped along a line perpendicular to the fuselage.

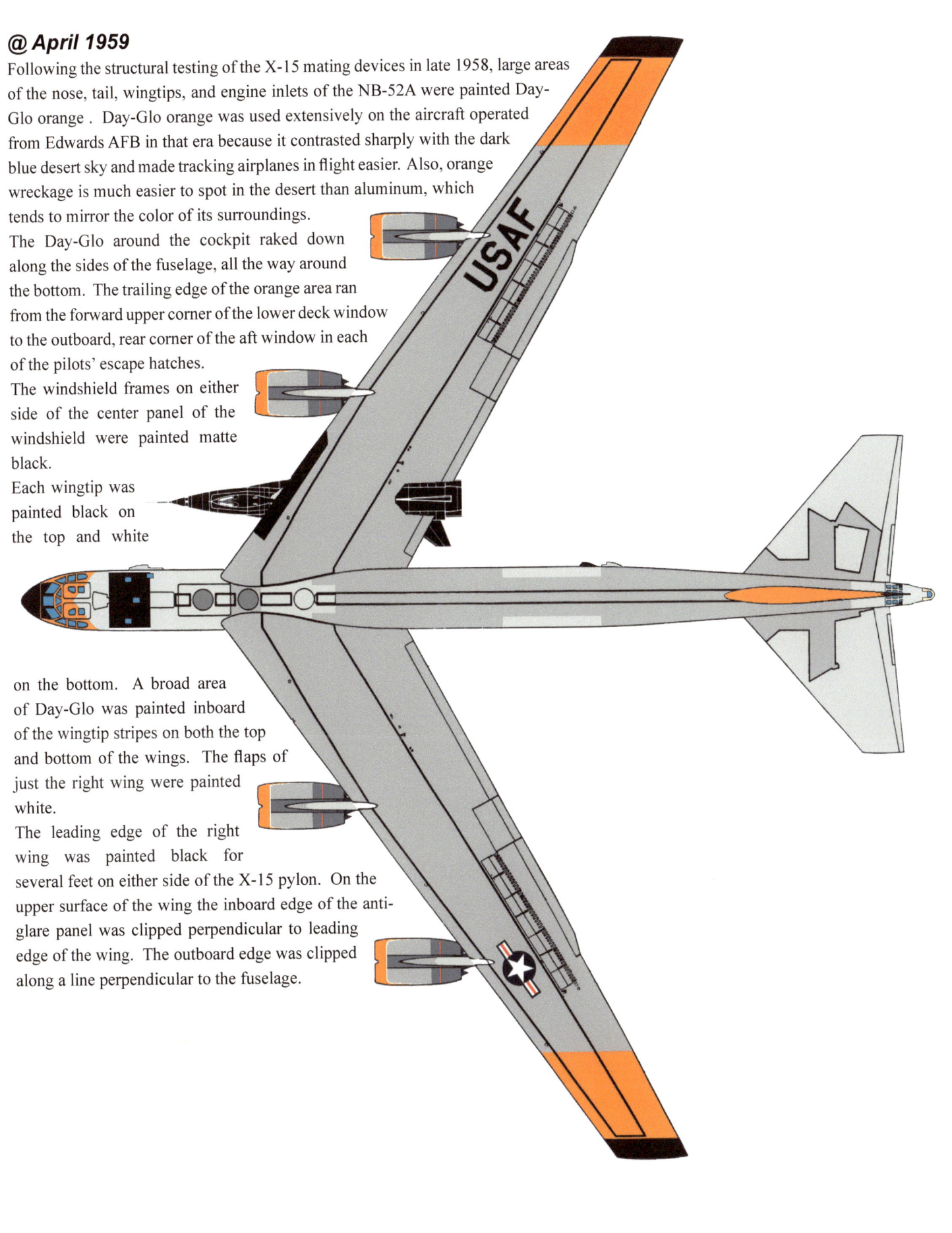

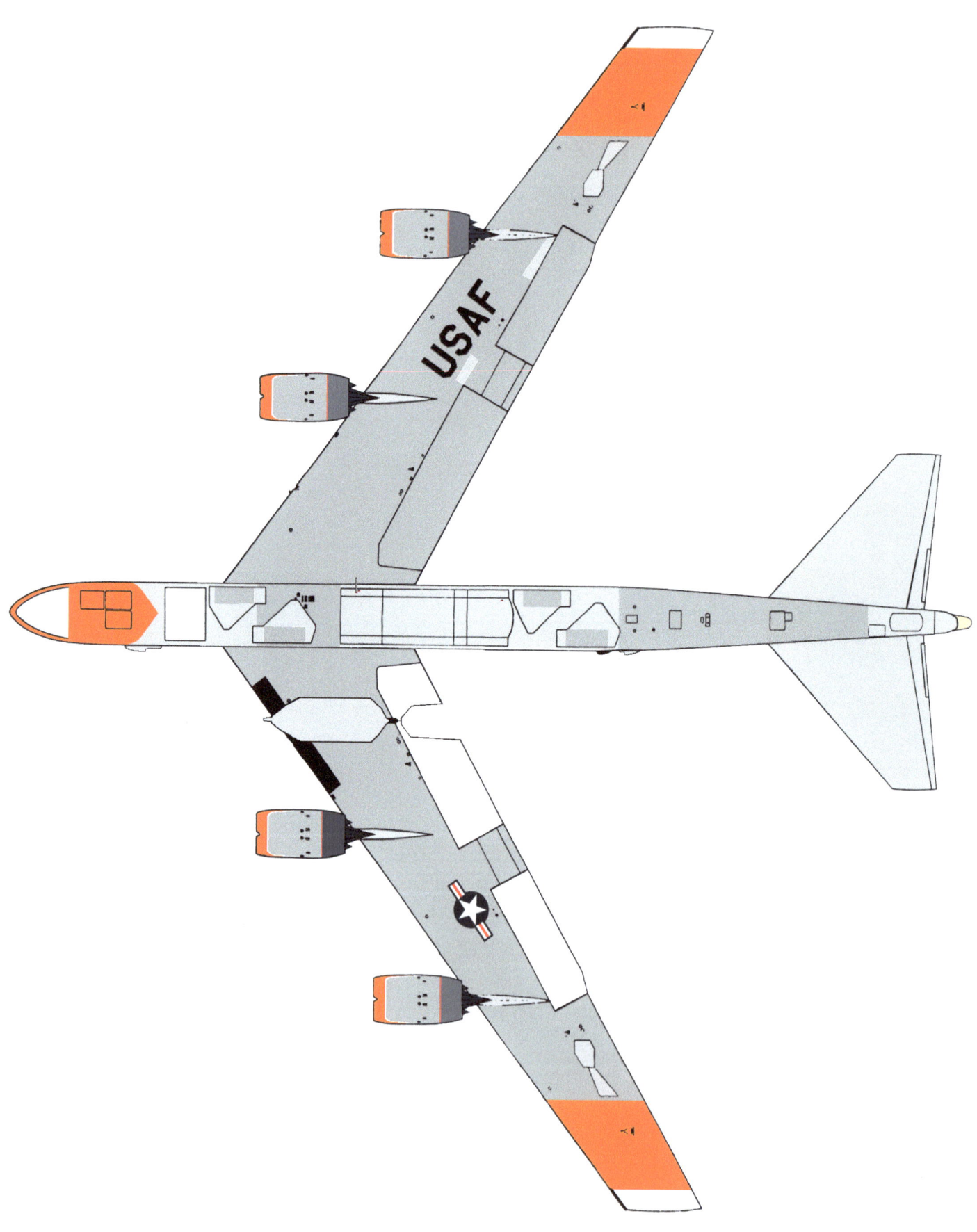

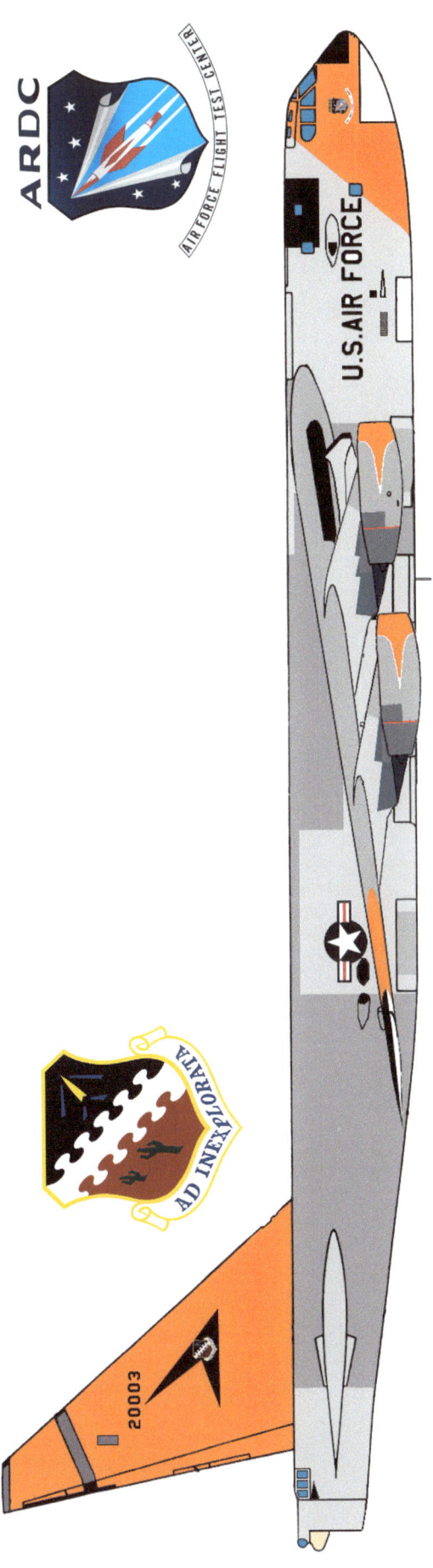
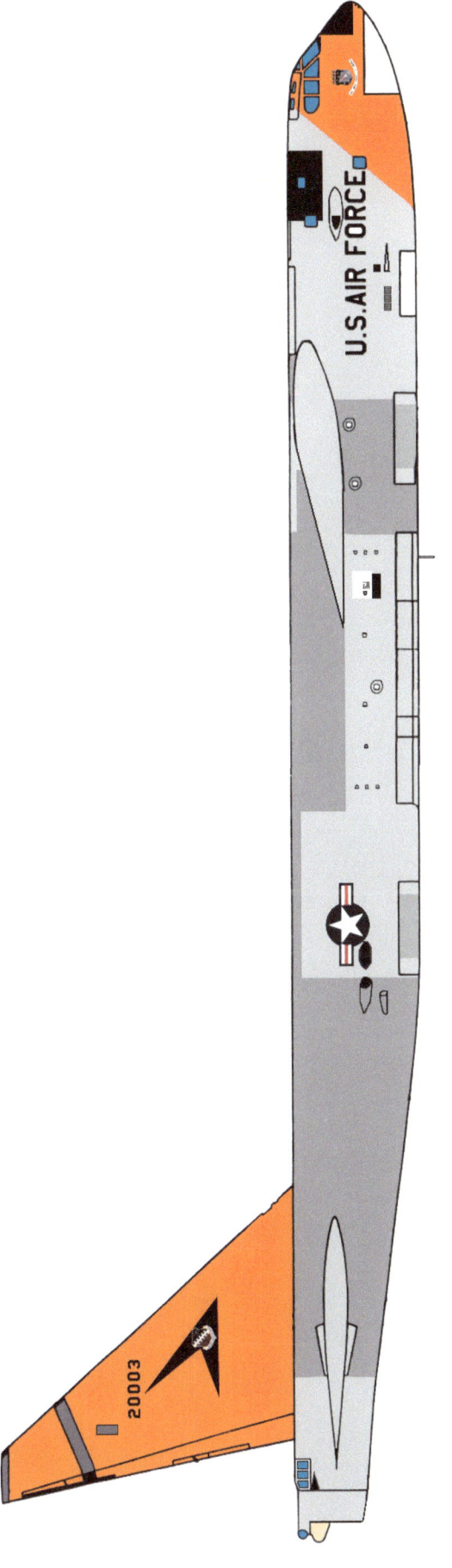

The Day-Glo on the engine inlets was scalloped to a point on either side of the nacelles. White cheat lines ran along the edge of the Day-Glo on the inlets. Almost the entire vertical stabilizer was Day-Glo orange. The huge orange area of the B-52's tail was a landmark that stood out distinctly from a great distance away. Spectators at the runway could see the tail of the B-52 approaching long before the rest of the bomber and its convoy of support vehicles came into view.
The 12-inch words "UNITED STATES AIR FORCE" on the forward fuselage were replaced with "U. S. AIR FORCE" in twenty-four inch tall letters.
A small black square was added directly above the NACA inlet near the belly radome.
Each side of the orange vertical stabilizer was adorned with a large black chevron. This was typical of the aircraft assigned to the AFFTC. The chevrons were not exact mirror images of one another. The AFFTC badge was applied within the black chevron on the tail.
The badge of the Air Research and Development Command (ARDC) was applied to the side of the fuselage, below the cockpit, where the AFFTC badge had been before.

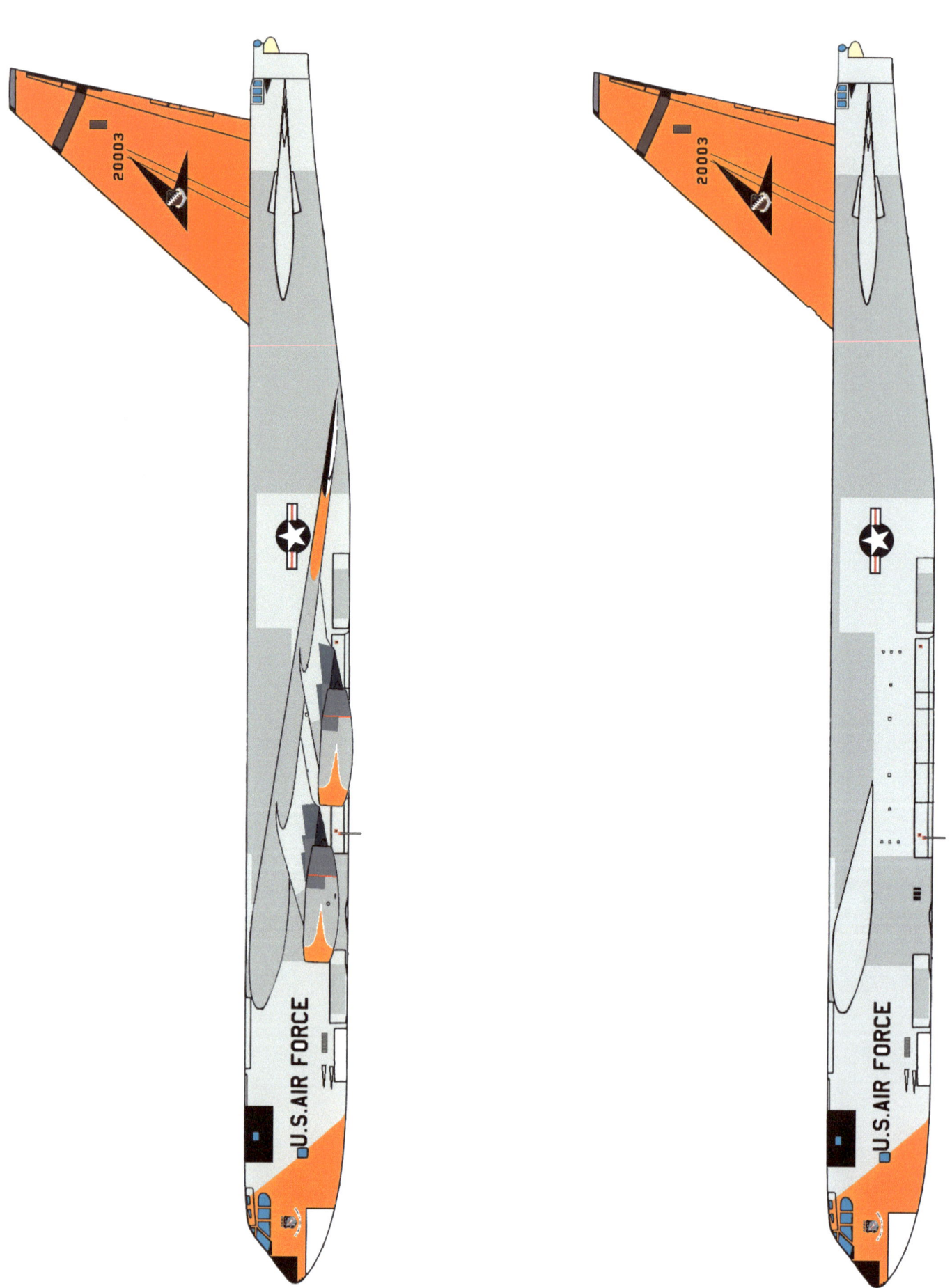

@ May 1959

Following the third captive flight on April 10, 1959 and before the May 19 Open House Display, the Day-Glo orange under the fuselage of the NB-52A was removed. The lower edge of the paint was a horizontal line about a foot and a half below the top of the chin radome. It ran back to a vertical edge that ran up to the center of the bottom of the lower deck window.

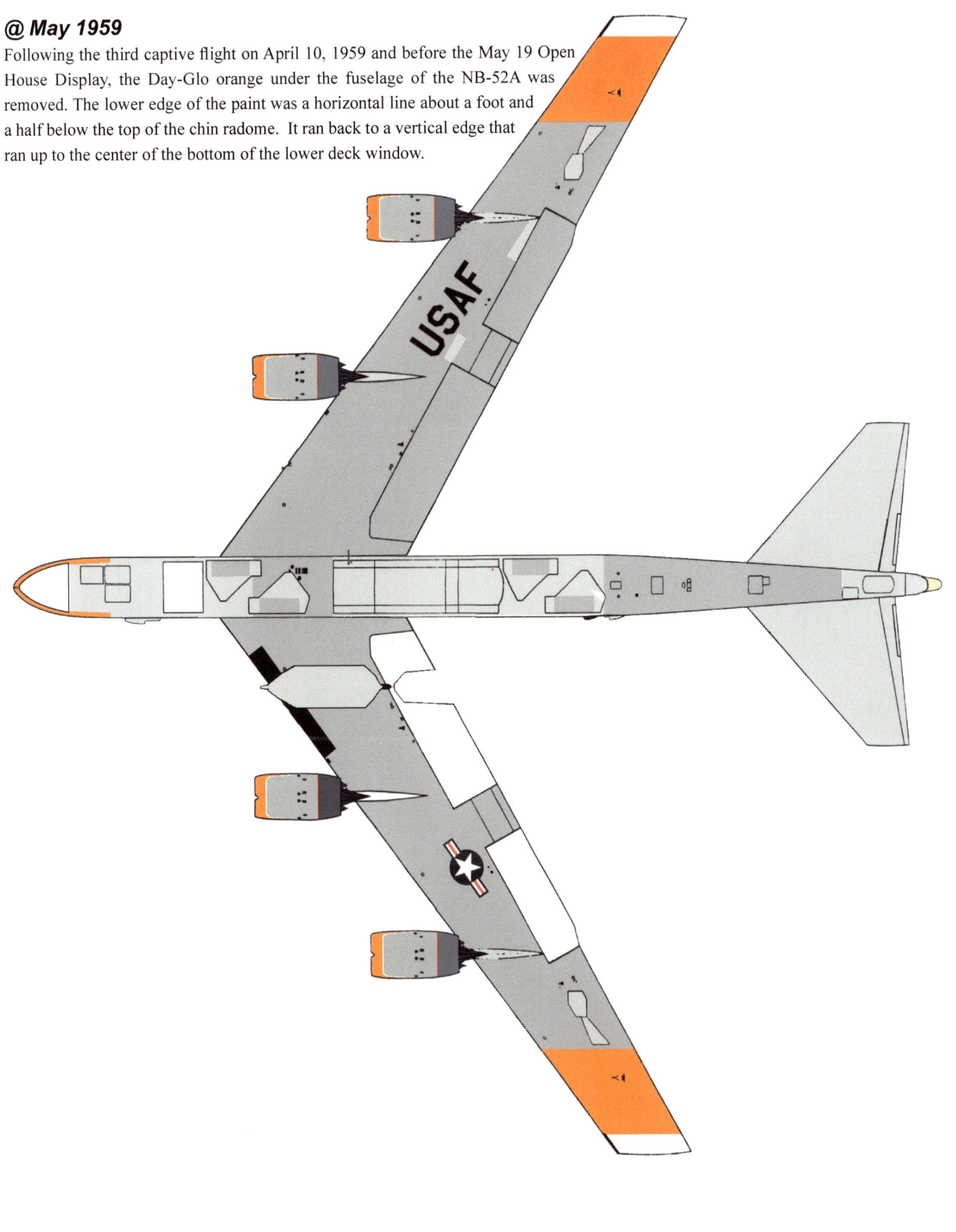

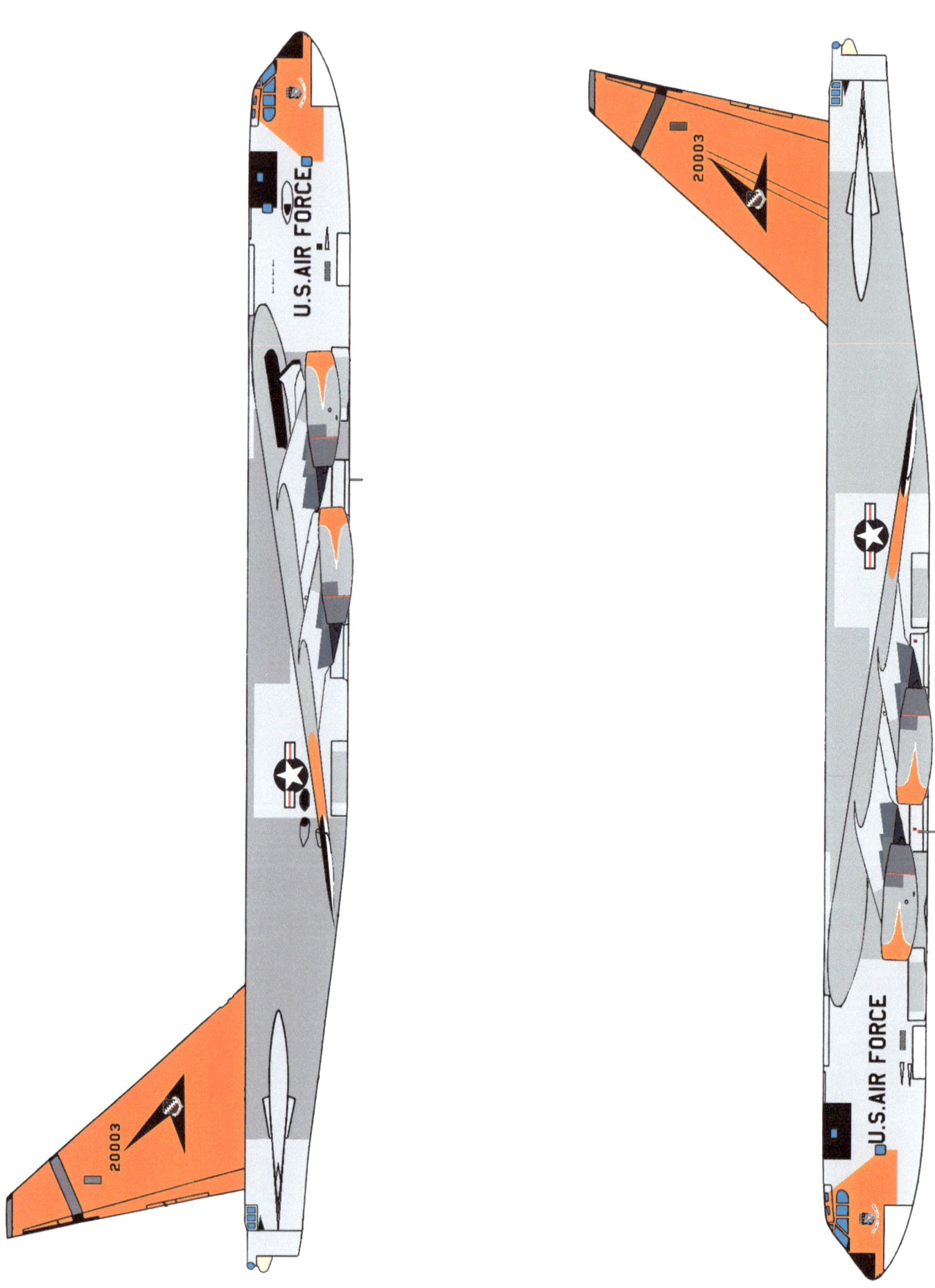

@ August 1959

A Plexiglas astrodome was installed at the location of the launch panel operator's station after the first captive flight of the X-15-2 was flown on July 24, 1959. The astrodome replaced the window, which had been installed in 1958. This significantly improved the launch panel operator's view of the X-15. The astrodome was installed before any X-15 mission marks were applied to the fuselage. At least one aborted X-15 mission was flown in September after the astrodome was added before X-15 mission marks began to be applied to the right side of the fuselage of the NB-52A.

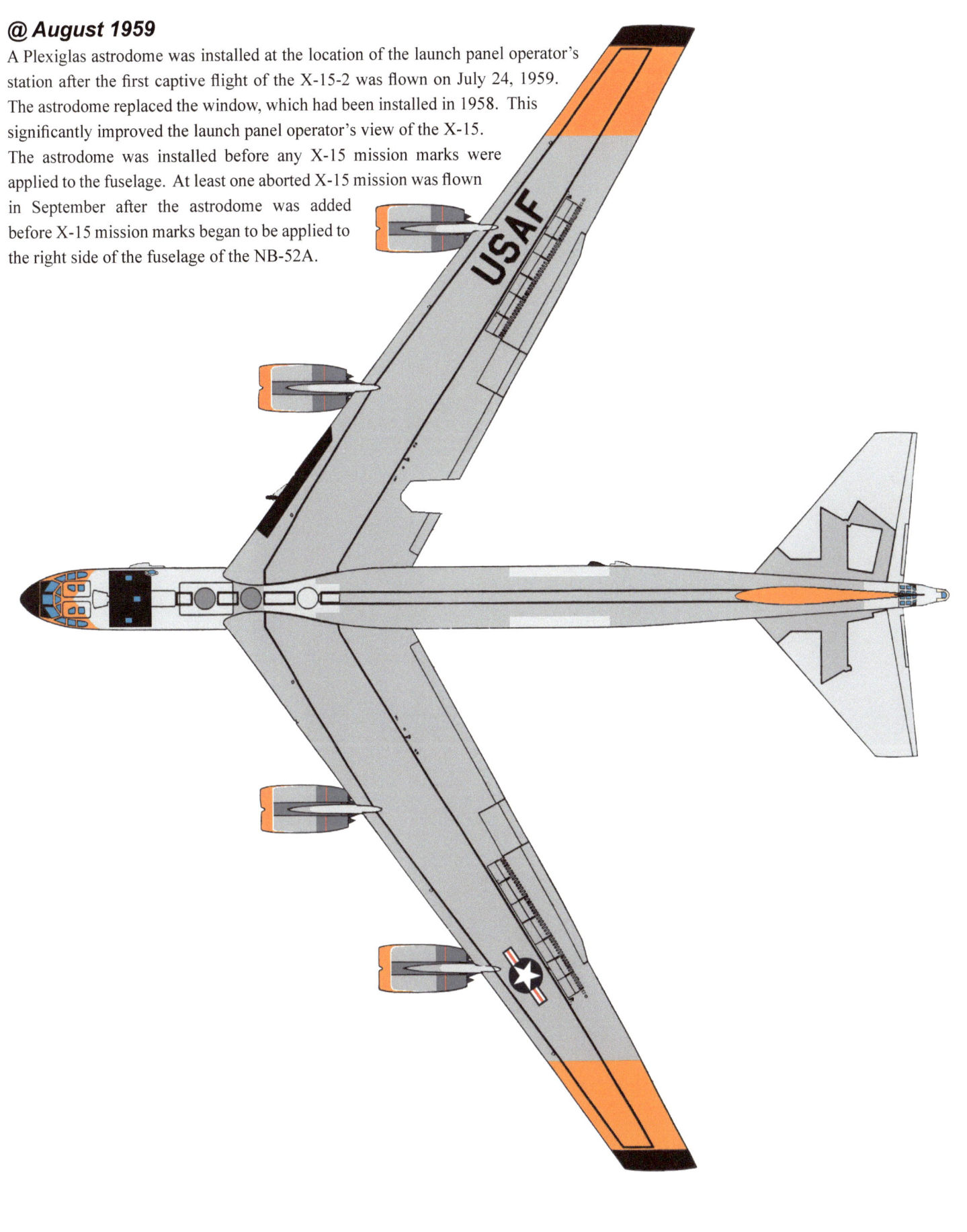

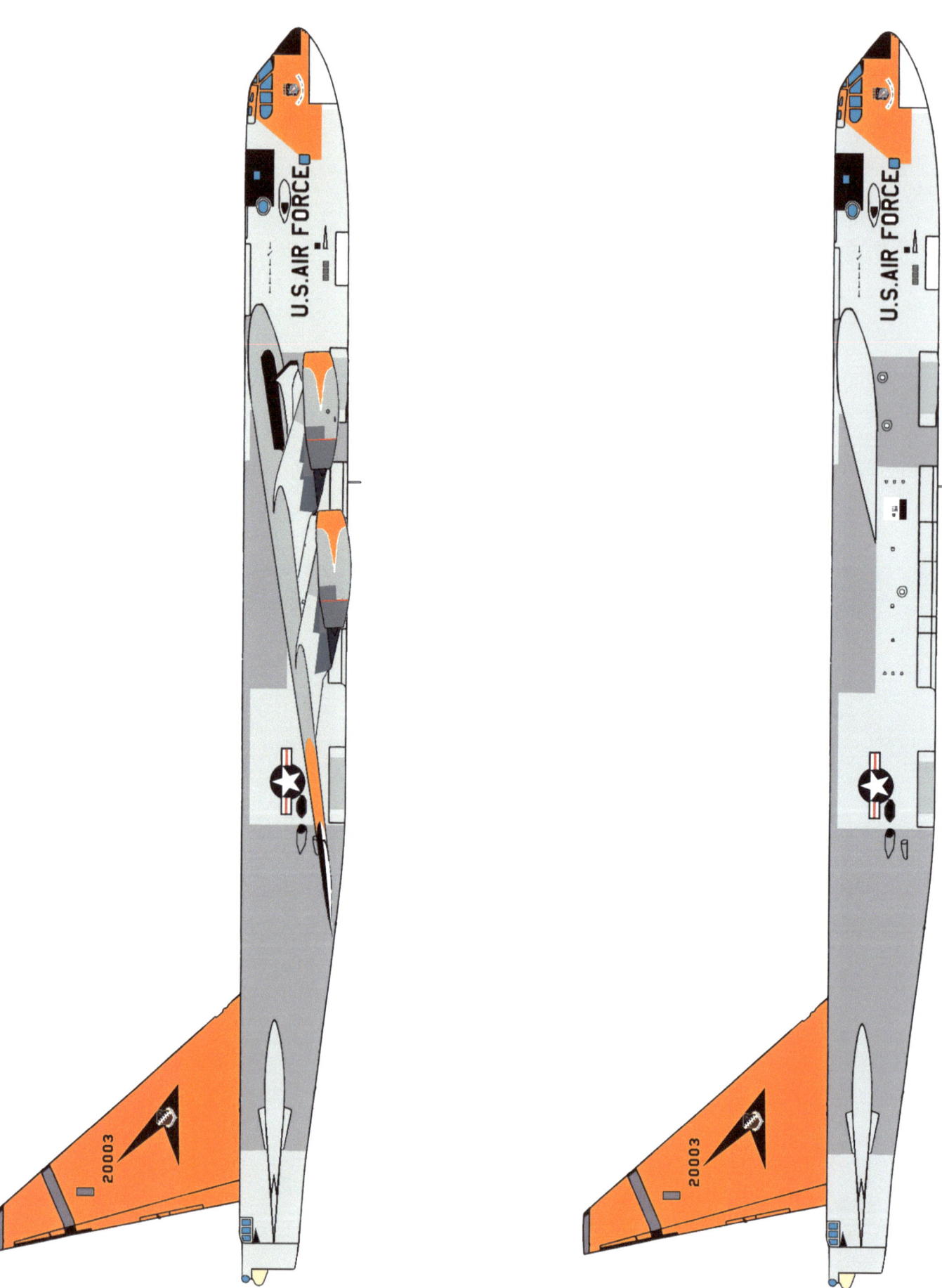

@ April 1960

The matte black area over the launch panel operator's station of the NB-52A was replaced with a more extensive area of semi-gloss white paint before the fourth flight of the X-15-1 was flown on April 13, 1960. The white area extended forward to the edge of the Day-Glo orange above the windshield.

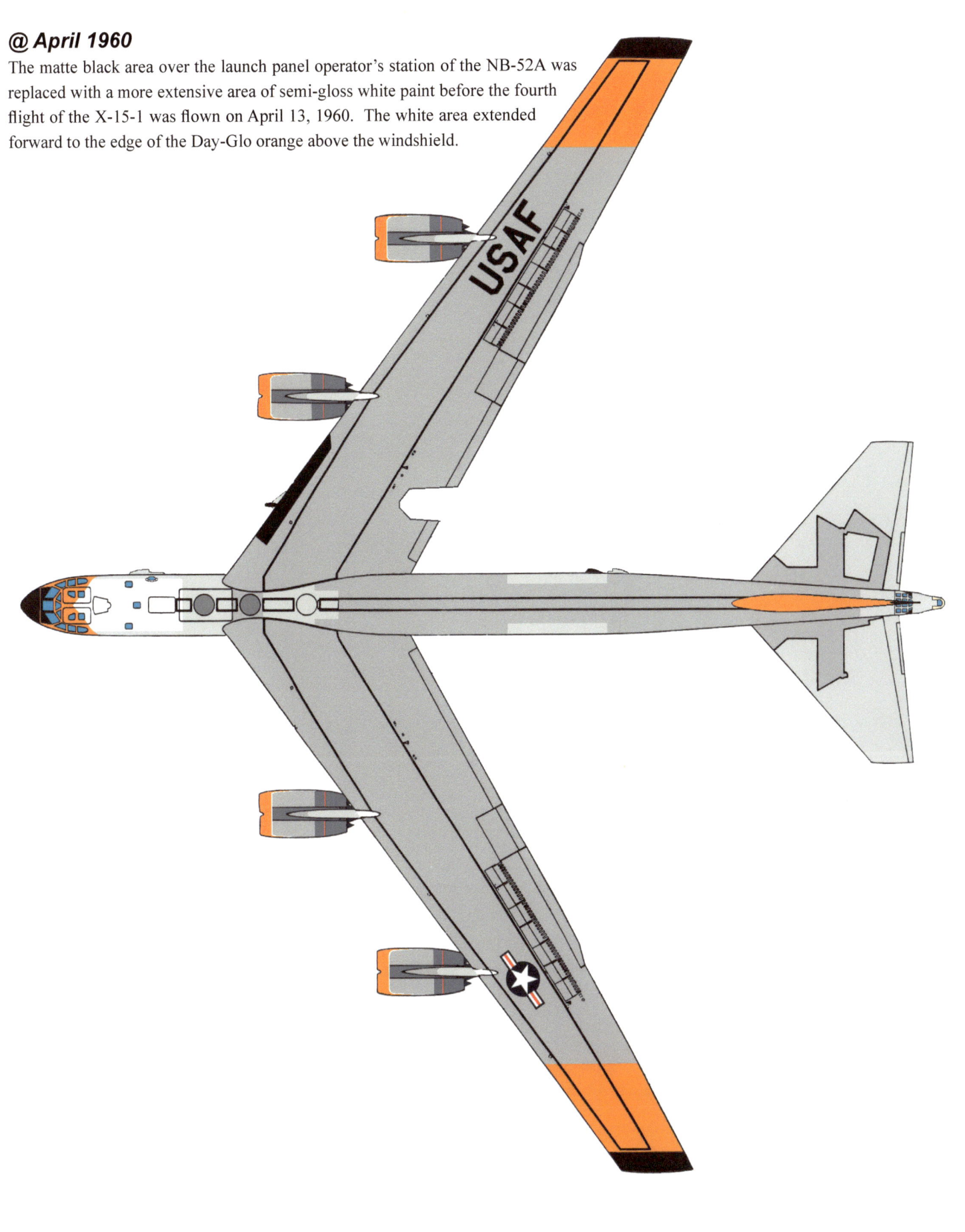

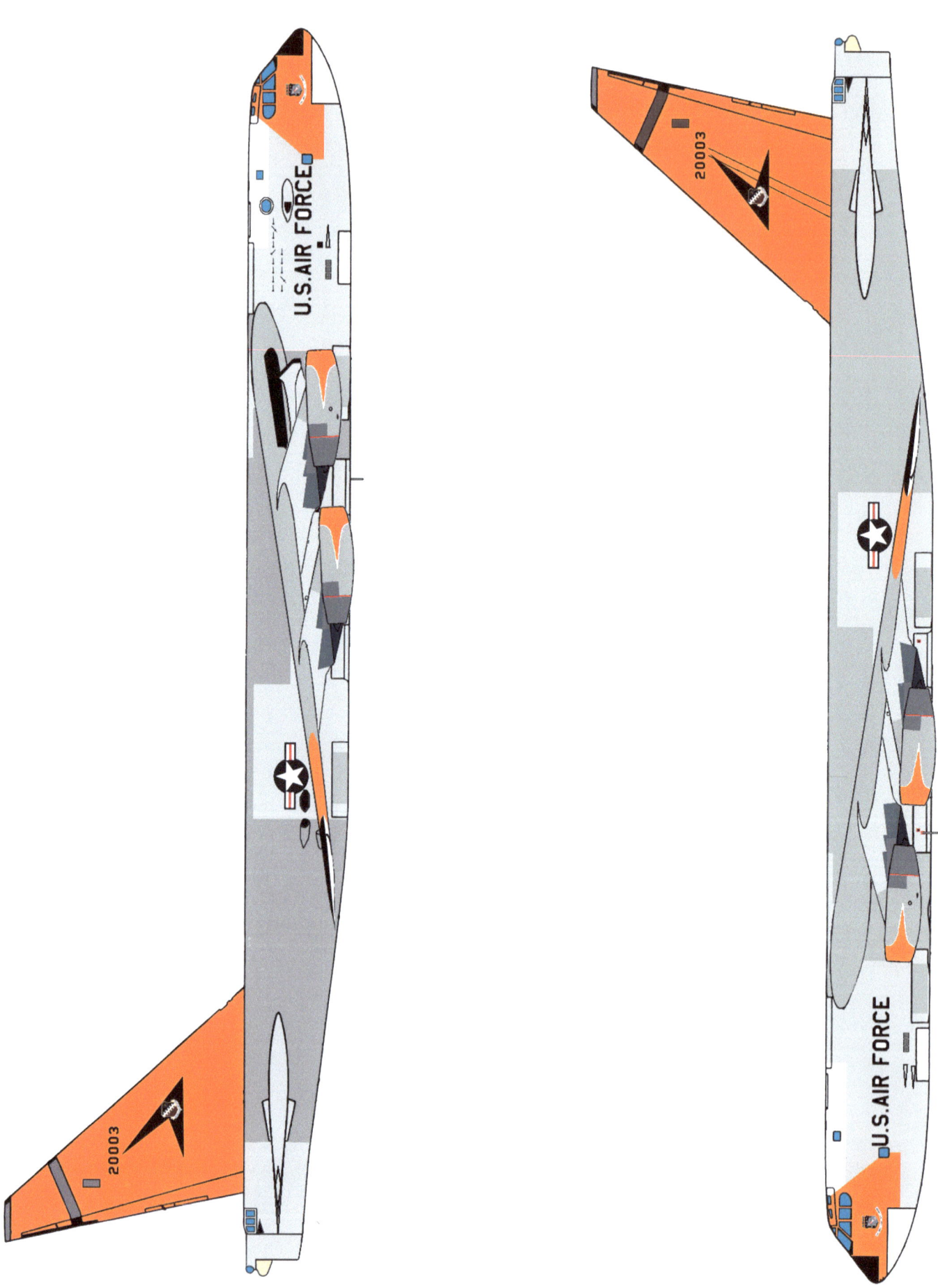

@ August 1960

A six-foot wide Day-Glo orange band was added to the rear fuselage. The trailing edge of this band was aft of the leading edge of the vertical stabilizer and forward of the leading edge of the horizontal stabilizer.

The lower edge of the Day-Glo orange around the cockpit of the NB-52A was moved up to a line even with the top of the chin radome.

The black anti-glare panel on the leading edge of the right wing was extended to reach from the fuselage to the inboard engine pylon.

The frames of the panes of the upper windshield were painted matte black.

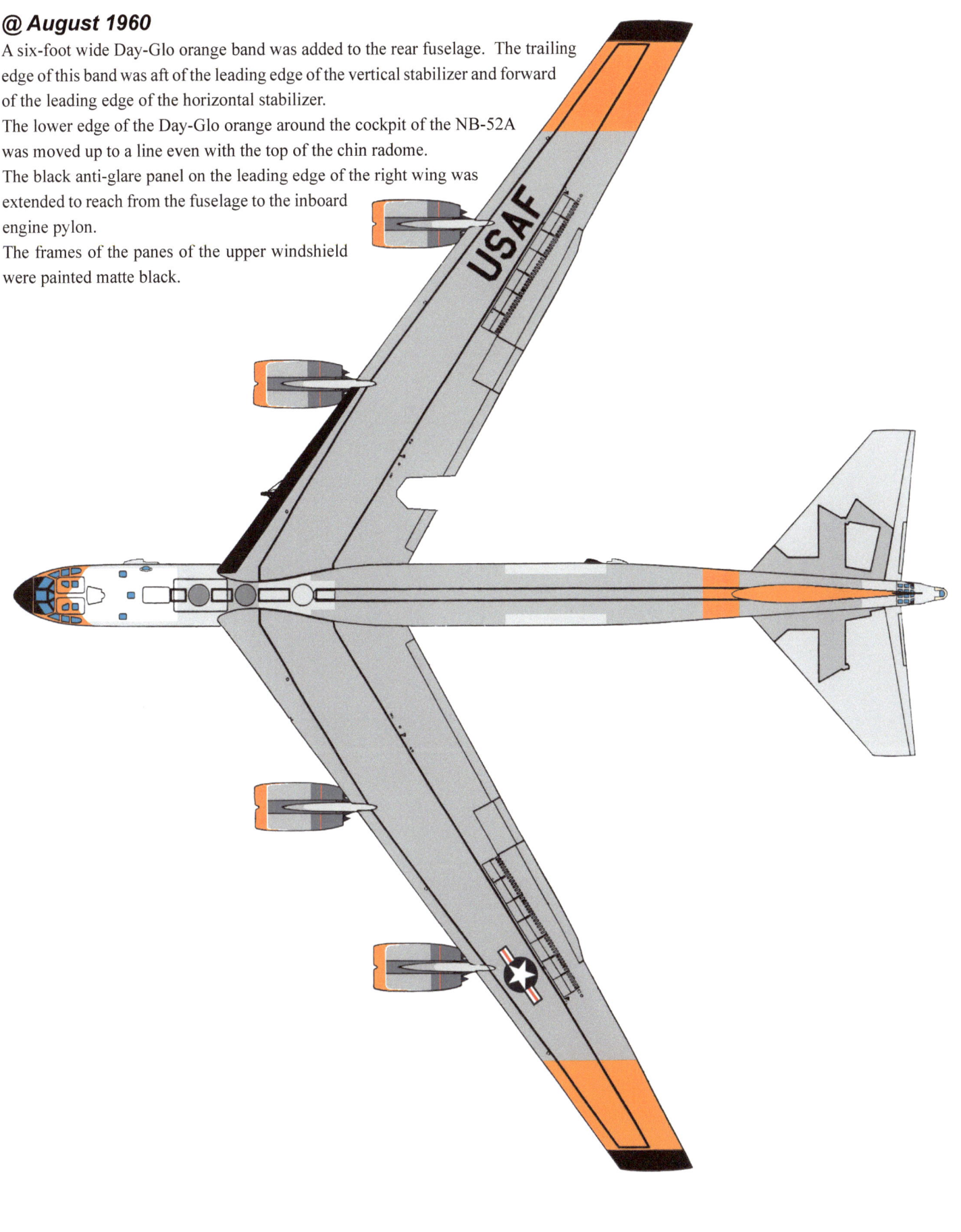

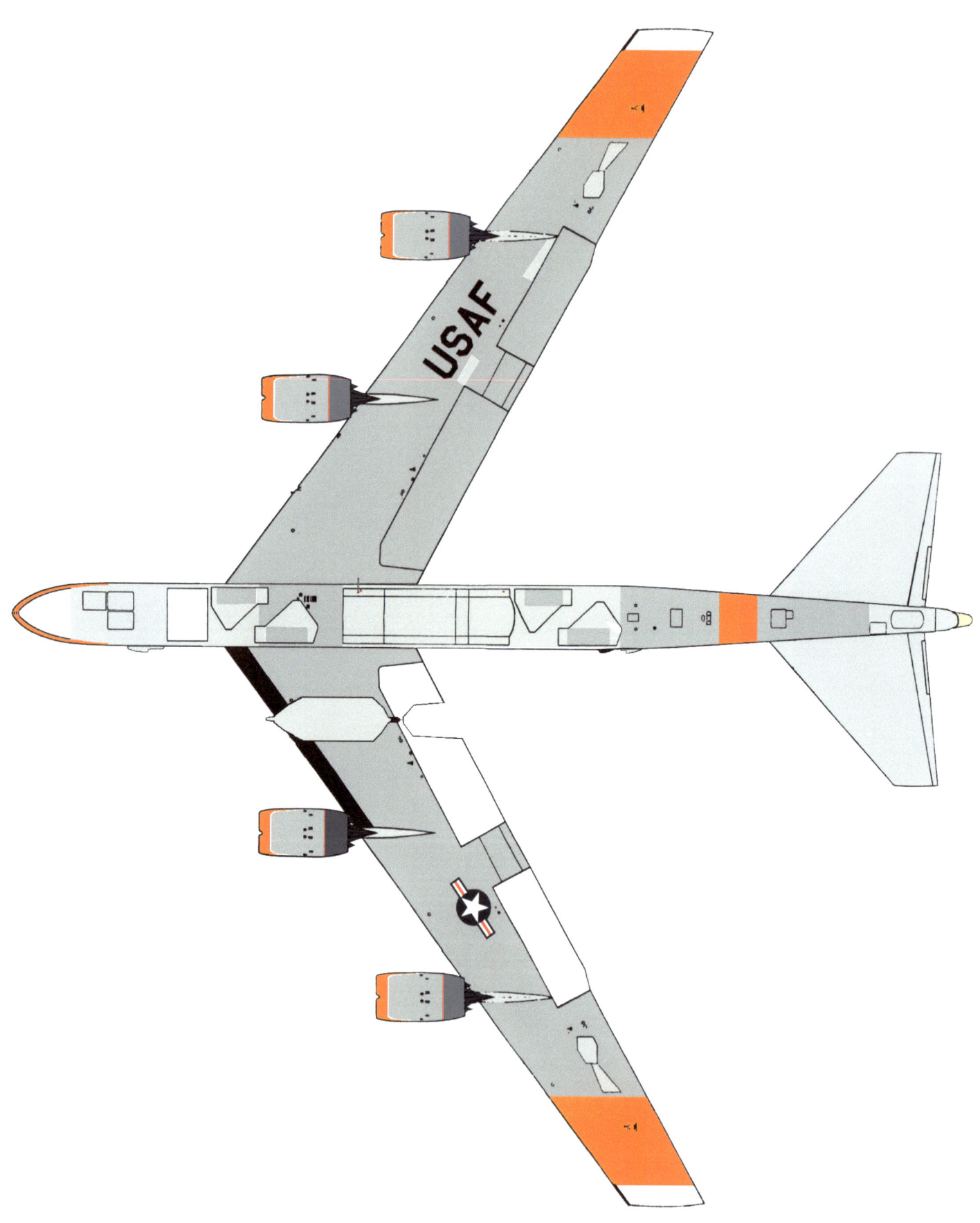

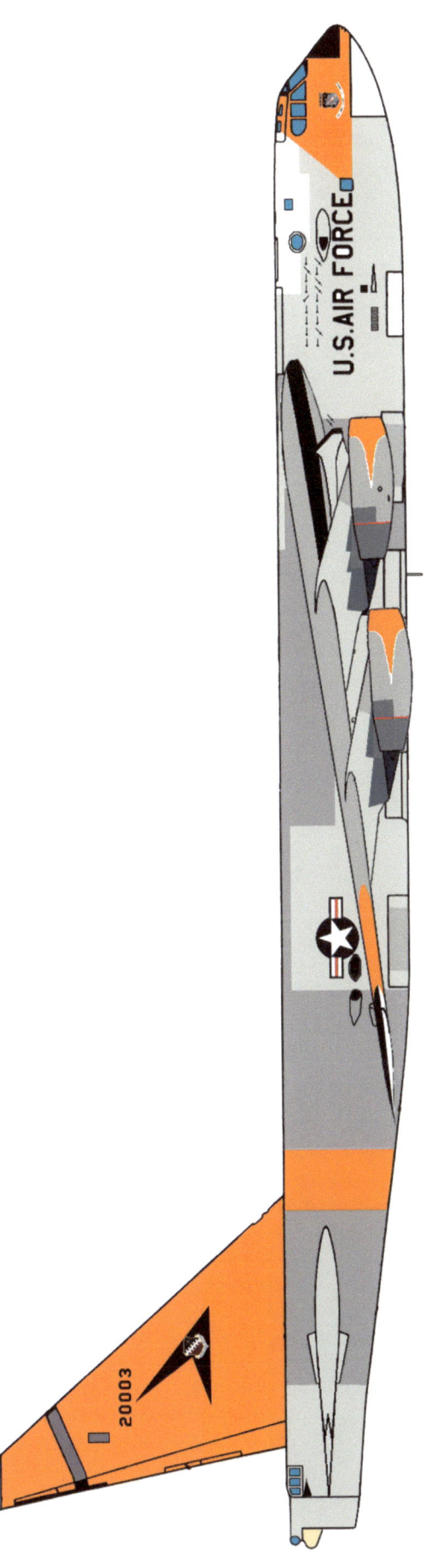

X-15 mission marks were added between the leading edge of the wing and the astrodome. Horizontal marks indicated captive carry and aborted missions. Upward angled marks denoted powered flights. The first glide flight was represented by a downward angled mark. The lower edge of the Day-Glo orange around the cockpit of the NB-52A was moved up to a line even with the top of the chin radome.

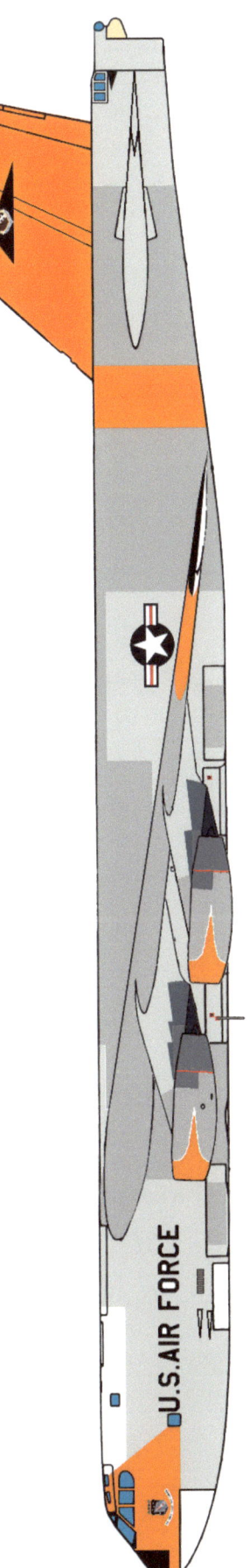

@ *October 1960*

The Day-Glo was stripped from the vertical stabilizer and engine inlets. The area of Day-Glo on the outer wings was reduced. The inboard edges moved outward and it no longer wrapped around the leading and trailing edges of the wing.

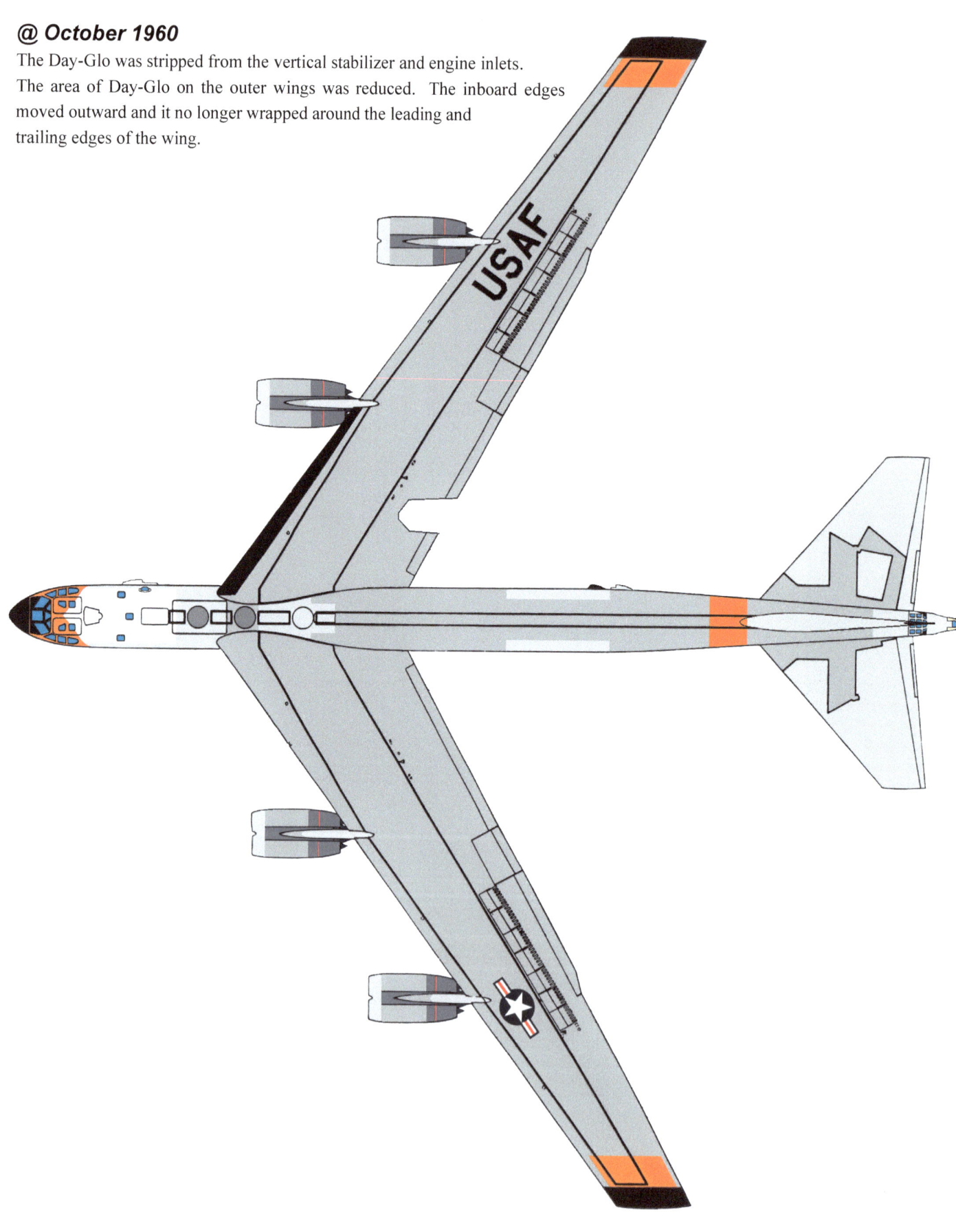

The metal exposed by the removal of some of the DayGlo orange from the underside of the wing was noticeably brighter than the surfaces that had not previously been painted.

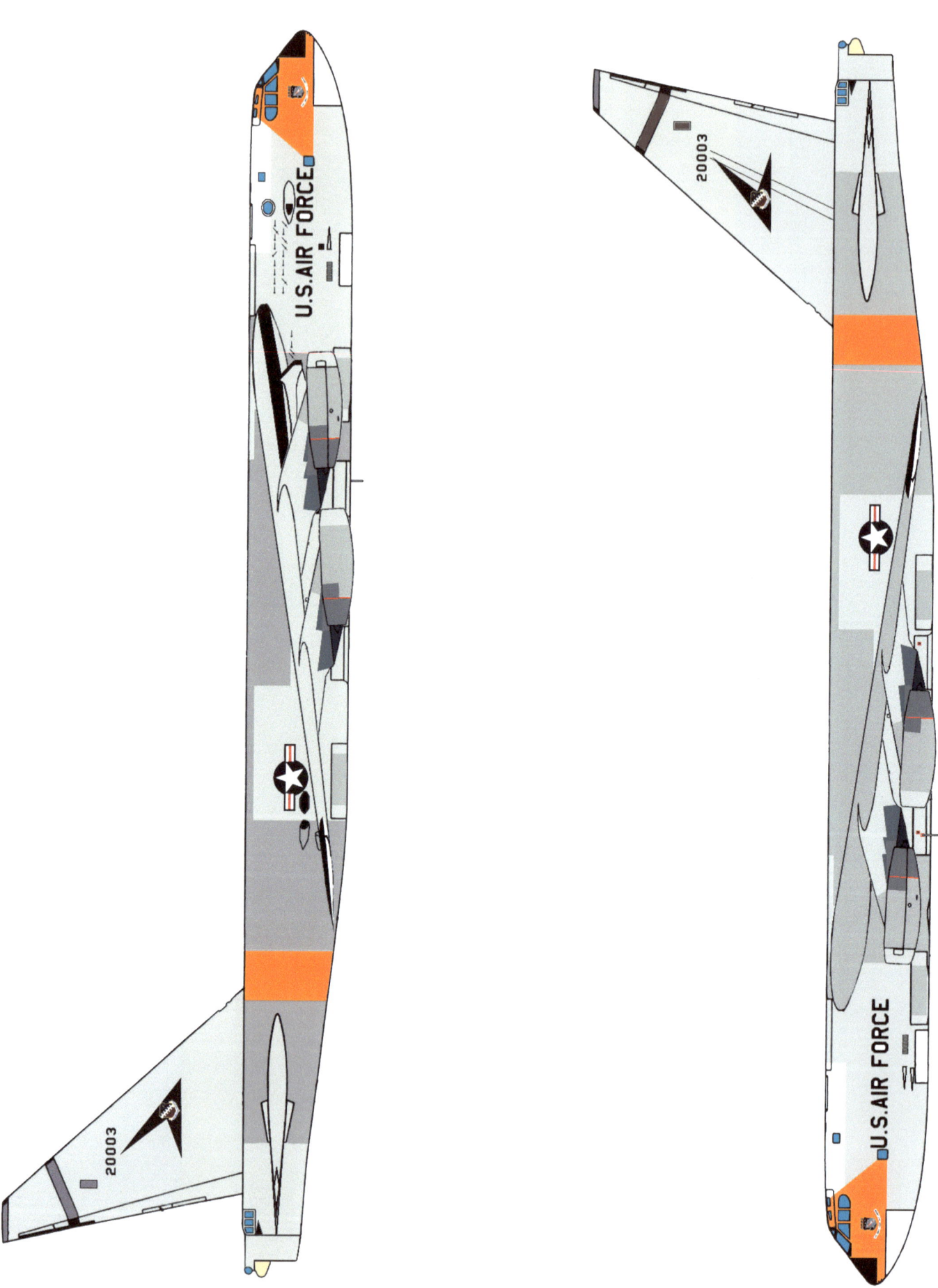

@ April 1961

The NB-52A was flown to Boeing's facilities at Wichita, Kansas in early 1961 for maintenance. While it was in the hands of Boeing, it was given a complete makeover. The first X-15 launch conducted after repainting was the fifteenth flight of the second X-15 on April 21, 1961.

The exposed metal areas of the NB-52A were painted silver.

The white area above the cockpit was extended forward all the way to the windshield.

The flaps on the left wing were painted white to match the flaps on the right wing.

The black anti-glare panel on the tail turret was restored.

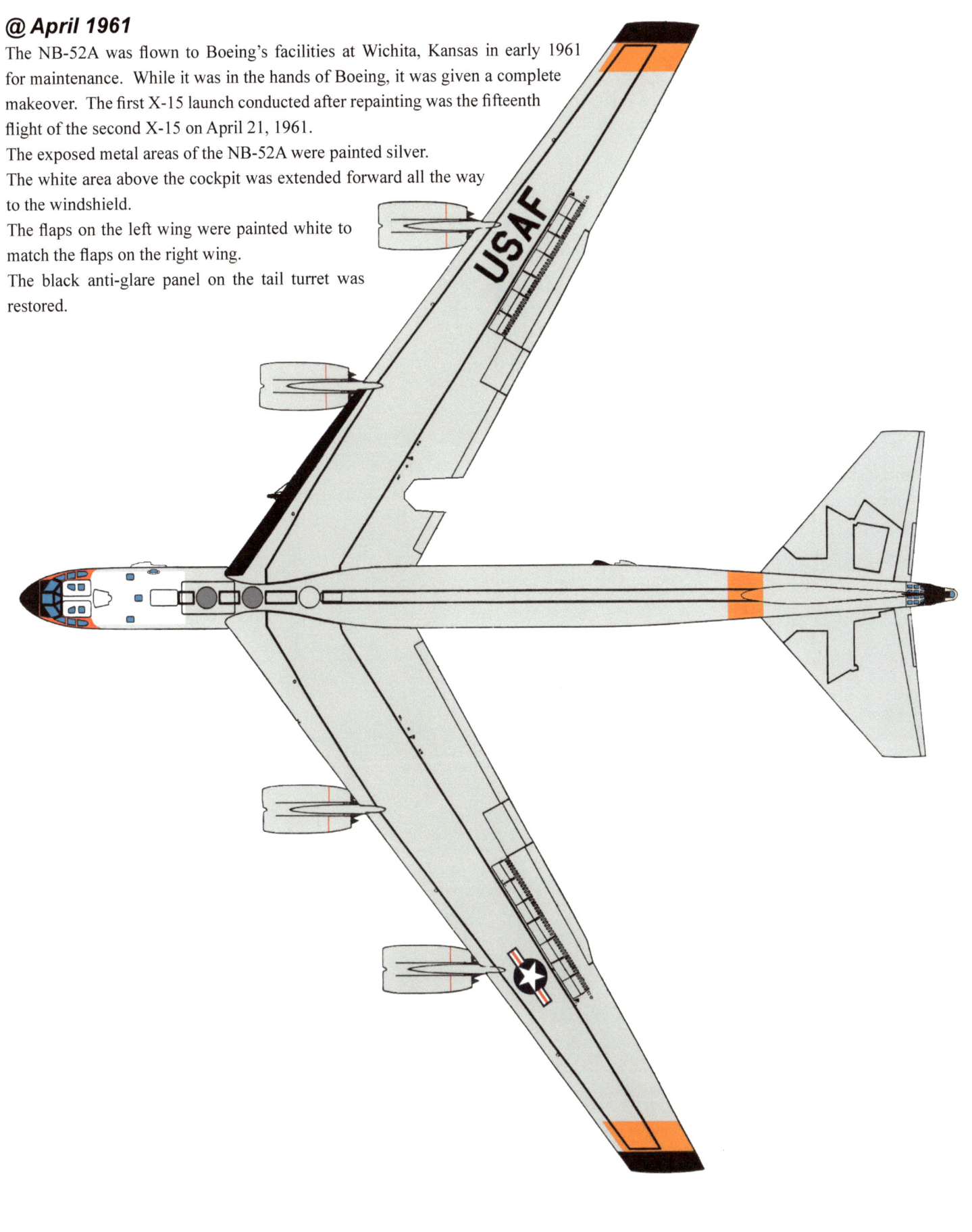

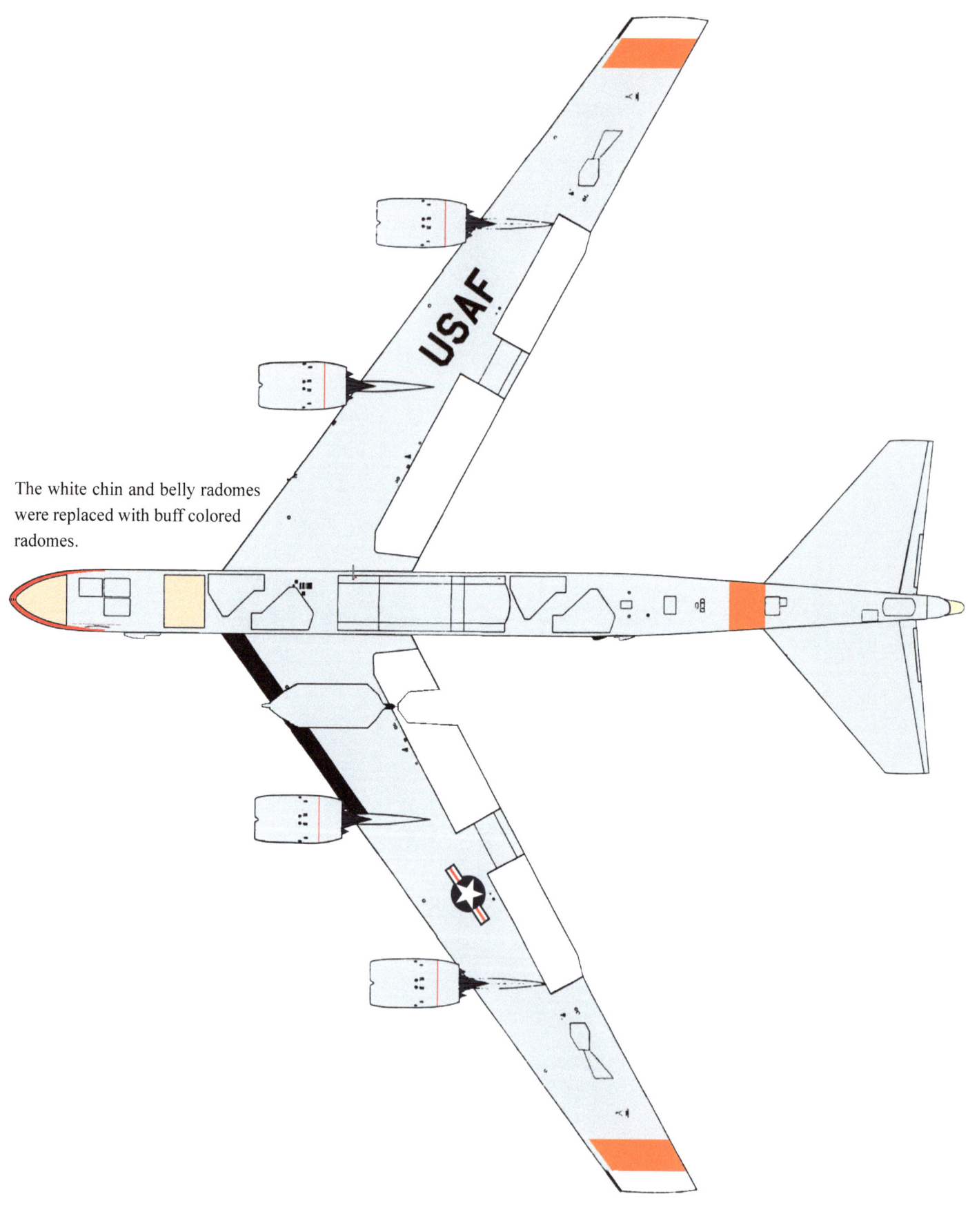

The white chin and belly radomes were replaced with buff colored radomes.

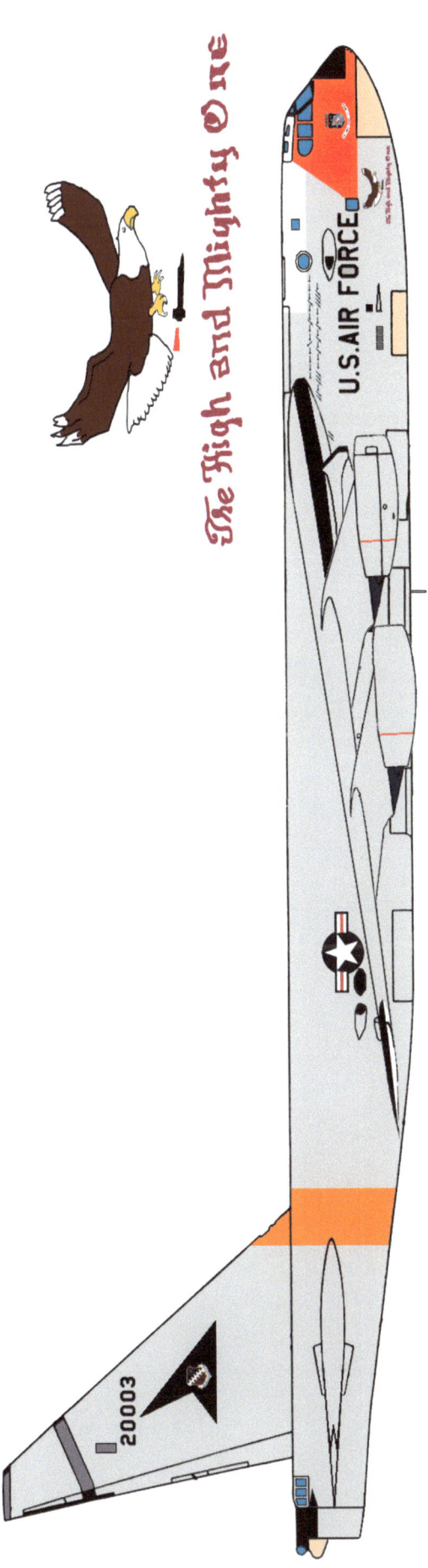
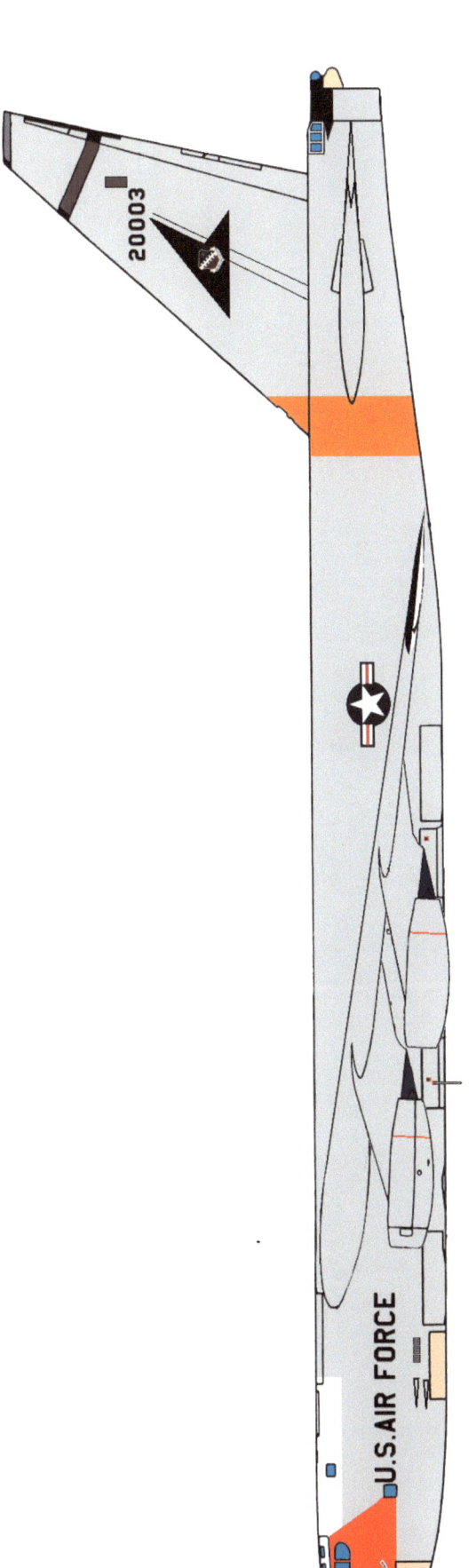

The X-15 mission marks were moved closer to the leading edge of the wing to make room for more.
The Day-Glo orange around the cockpit of the NB-52A was replaced with Day-Glo red. The lower edge of the Day-Glo as brought up to the level of the top of the chin radome, and Day-Glo no longer extended over the top of the fuselage.
The Day-Glo band around the rear fuselage remained orange and was moved aft.   The rear end of the band was even with the leading edge of the horizontal stabilizer and wrapped around the tip of the leading edge of the vertical stabilizer.
The shape of the black chevrons on the vertical stabilizer was altered. The rear edge of the chevrons was moved aft to make the chevrons broader.
The X-15 mission marks were replaced with a more compact set.
The NB-52A acquired nose art featuring an eagle with an X-15 in its talons. The name under the eagle read "The High and Mighty One" in red letters.

@ *May 1962*

The Day-Glo orange band around the rear fuselage was replaced with a Day-Glo red band.

The buff colored belly radome was replaced with a white radome.

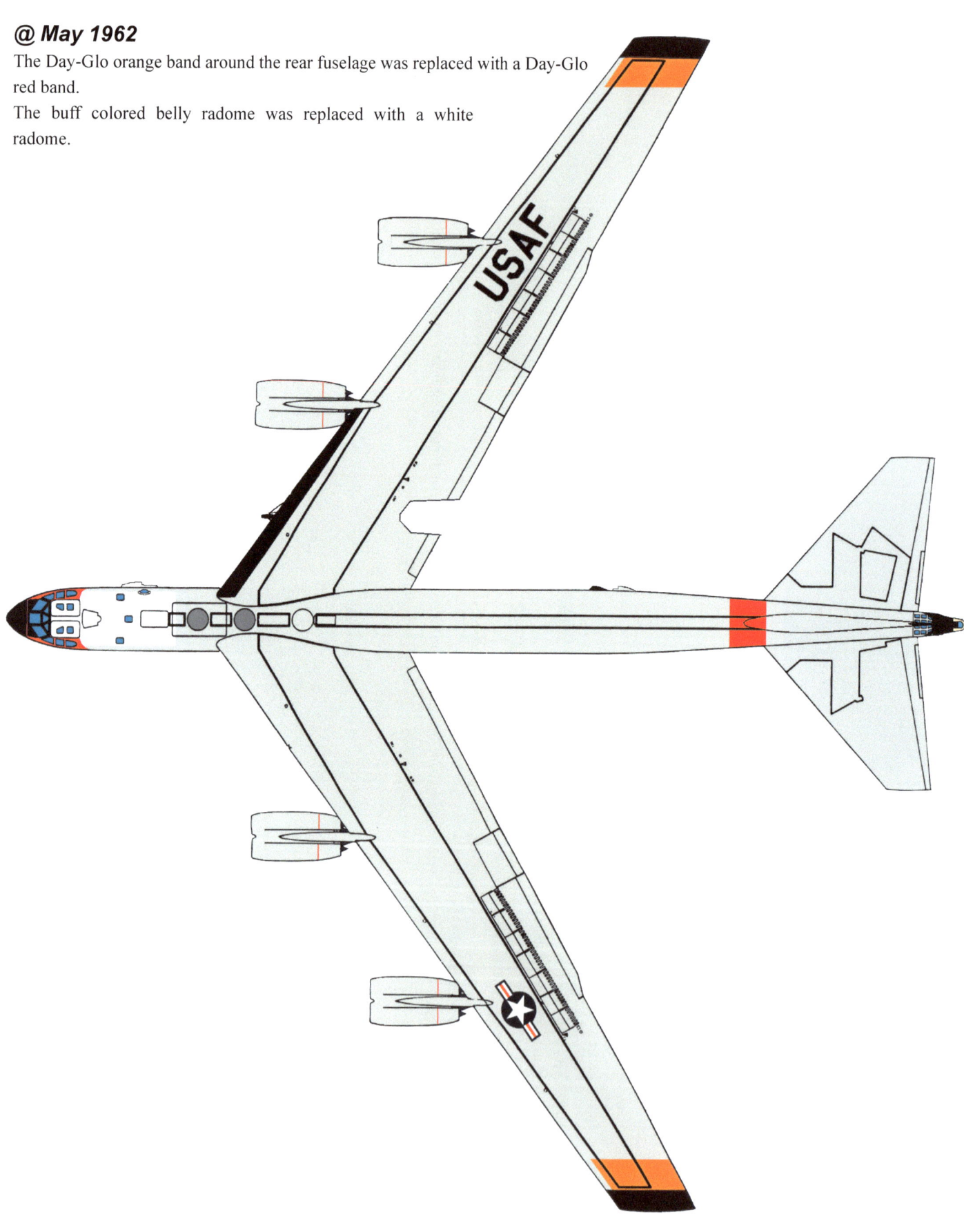

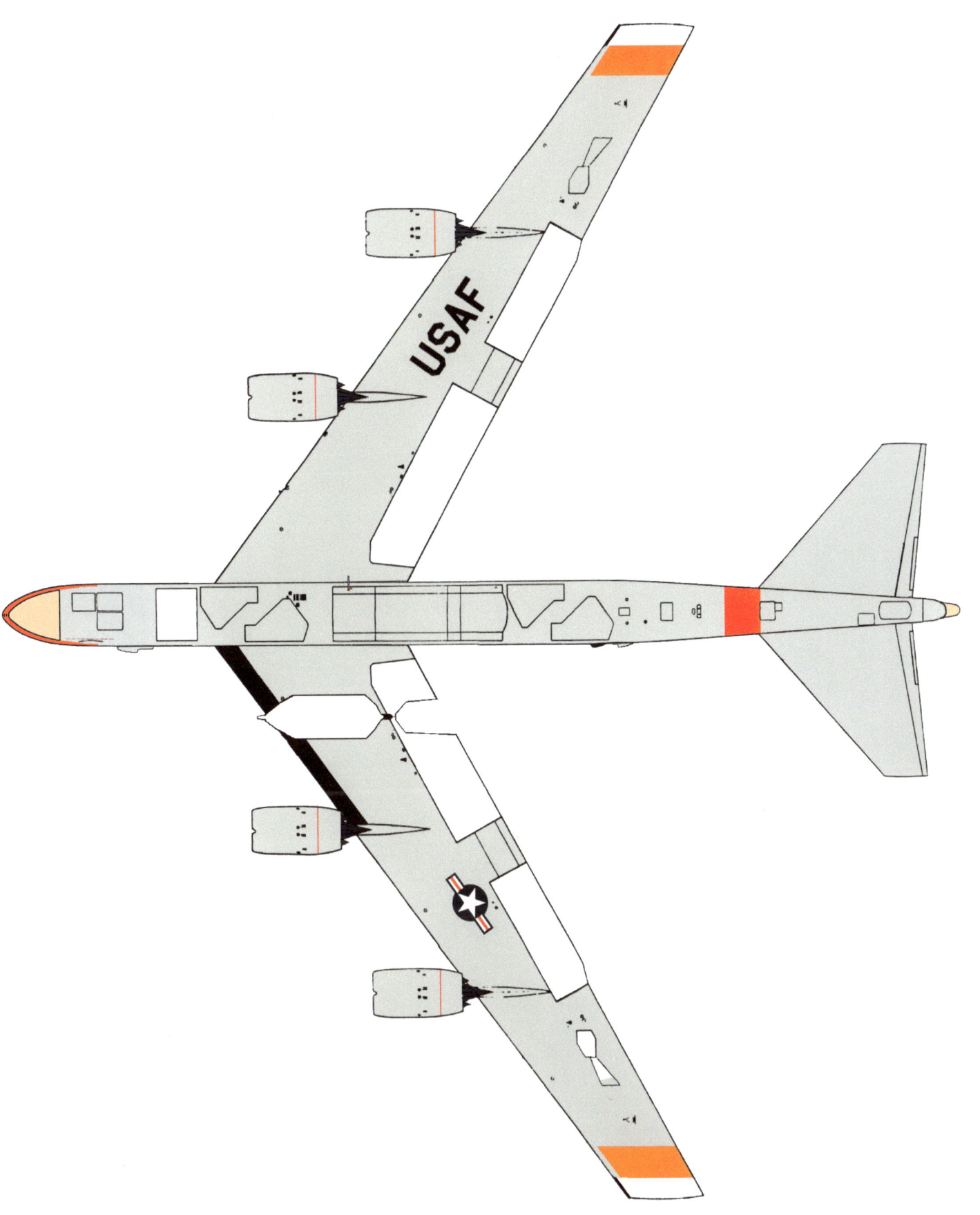

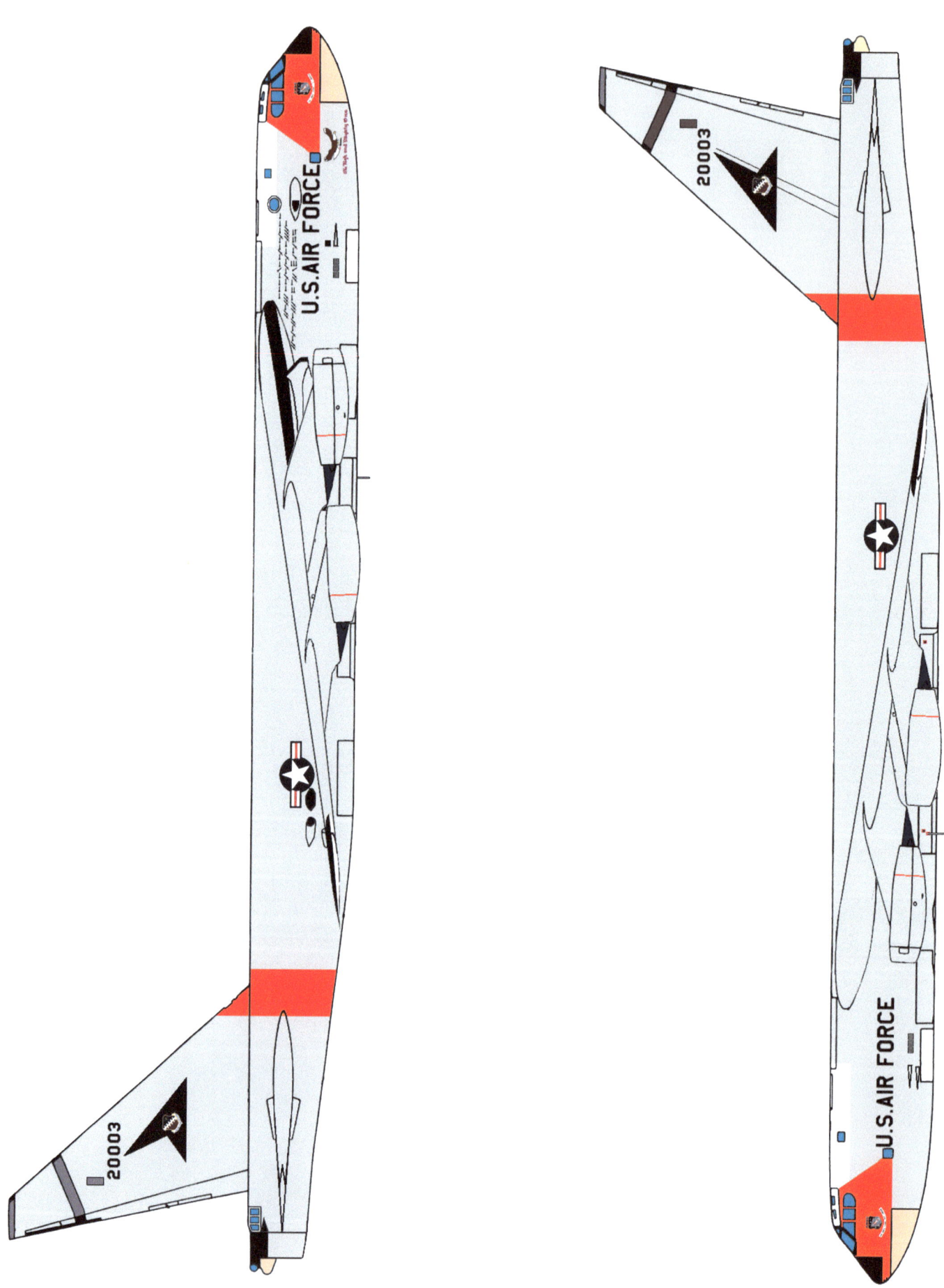

@ *May 1963*

The changes described below probably occurred while the NB-52A was at the Boeing facilities in Wichita, Kansas between December 1962 and April 1963.

The area above the windshield forward of the Day-Glo band was painted white.

The formerly black frames of the windshield were painted silver.

The Day-Glo orange at the wing tips was extended aft to the trailing edge of the wing.

The X-15 mission marks on the right fuselage of the NB-52A were removed and replaced in a more compact arrangement to make room for more marks.

The engine nacelles and the aft portions of the engine pylons were painted gloss light gray.

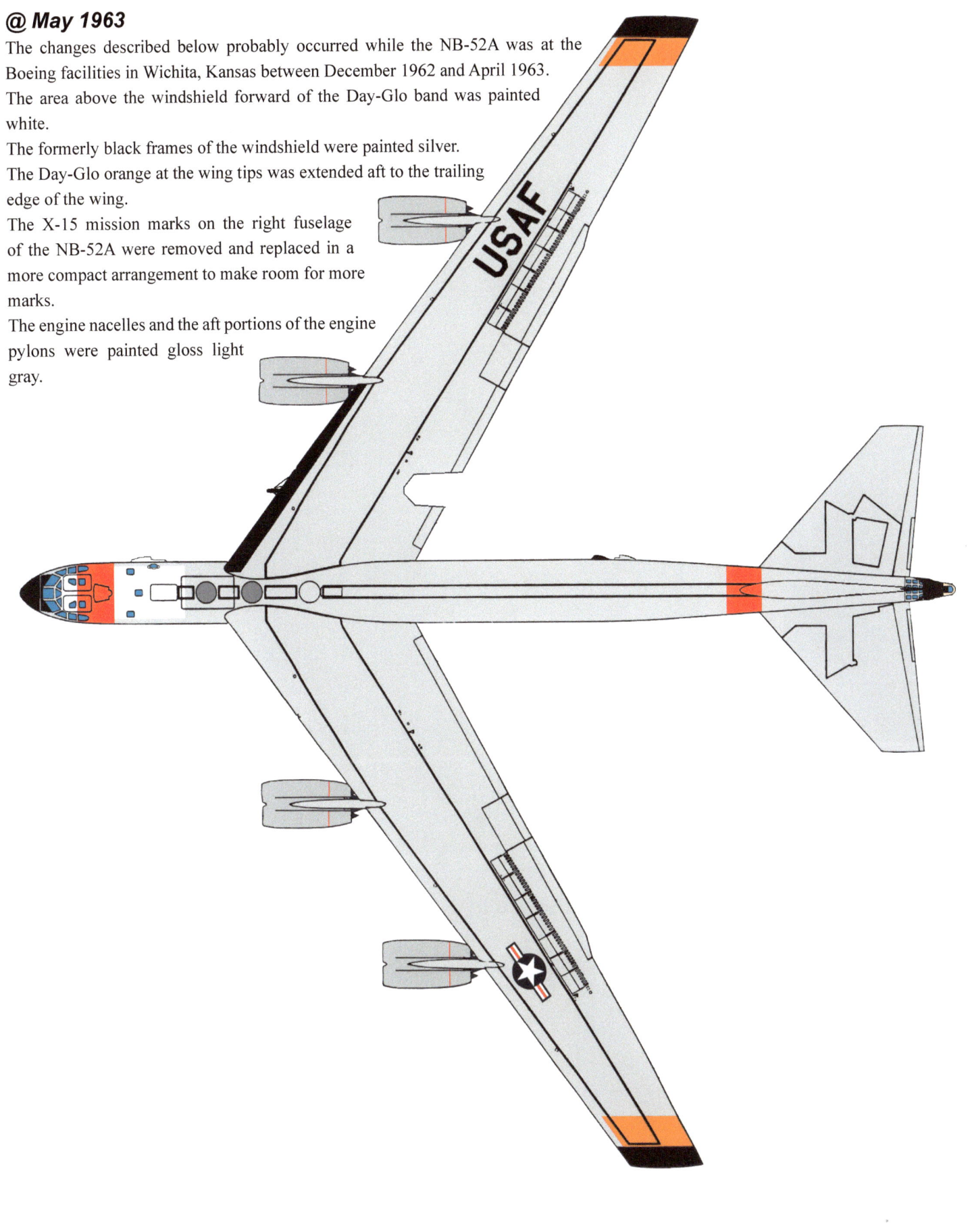

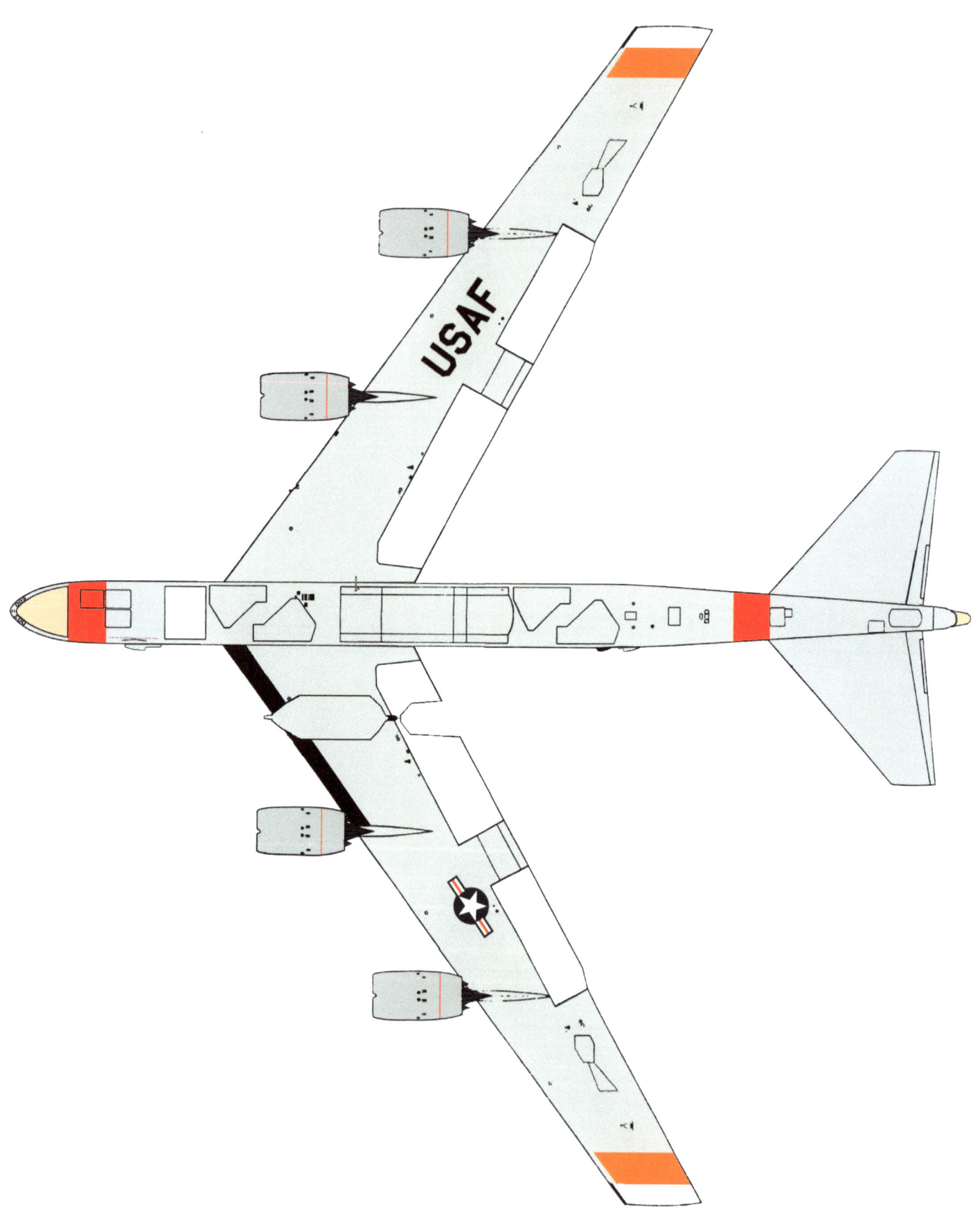

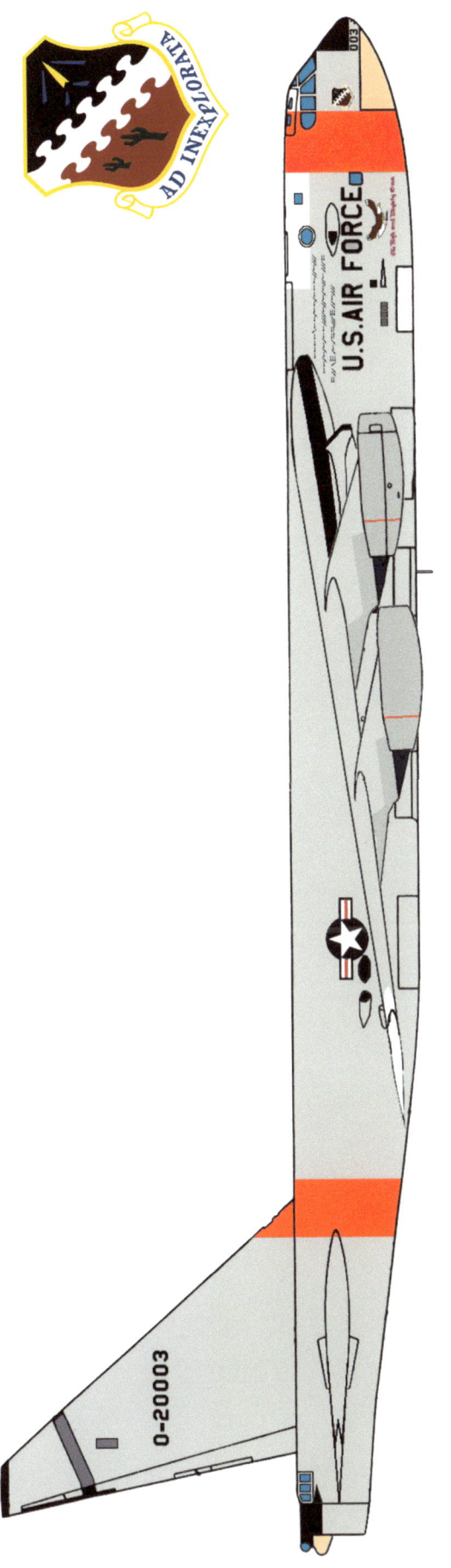
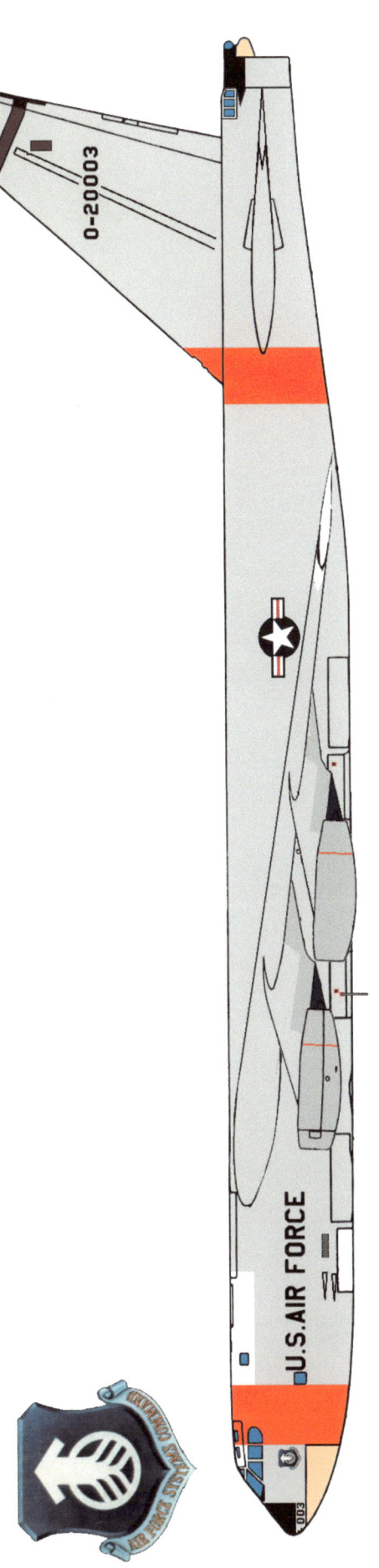

The X-15 mission marks were replaced with a more compact set.
The Day-Glo red on the forward fuselage of the NB-52A was replaced with a band that ran all the way around the fuselage behind the chin radome and in front of the lower deck window. Only the most rearward windowpane of the windshield extended into the Day-Glo area. The dividing line between the red and silver ran up the center of the frame immediately forward of this windowpane.
The eagle and the name "The High and Mighty One" were moved aft to a position under the word "FORCE".
The ARDC badges on the forward fuselage were replaced with the AFFTC badge on the right side and the Air Force Systems Command (AFSC) badge on the left side.
The black chevron was removed from the tail.
A zero was added to the beginning of the tail number to reflect the fact that the airplane was now over ten years old. The tail number was moved down from its previous position on the vertical stabilizer.
The last three digits of the tail number, "003", were repeated on either side of the nose between the nose radome and the chin radome.
The gray di-electric panel on the top of the vertical stabilizer was painted black or replaced with a black panel.

@ May 1965

The frames of the forward three panes of the windshield and the eyebrow windows were painted matte black. The formerly black top of the outboard wing tips was painted white.

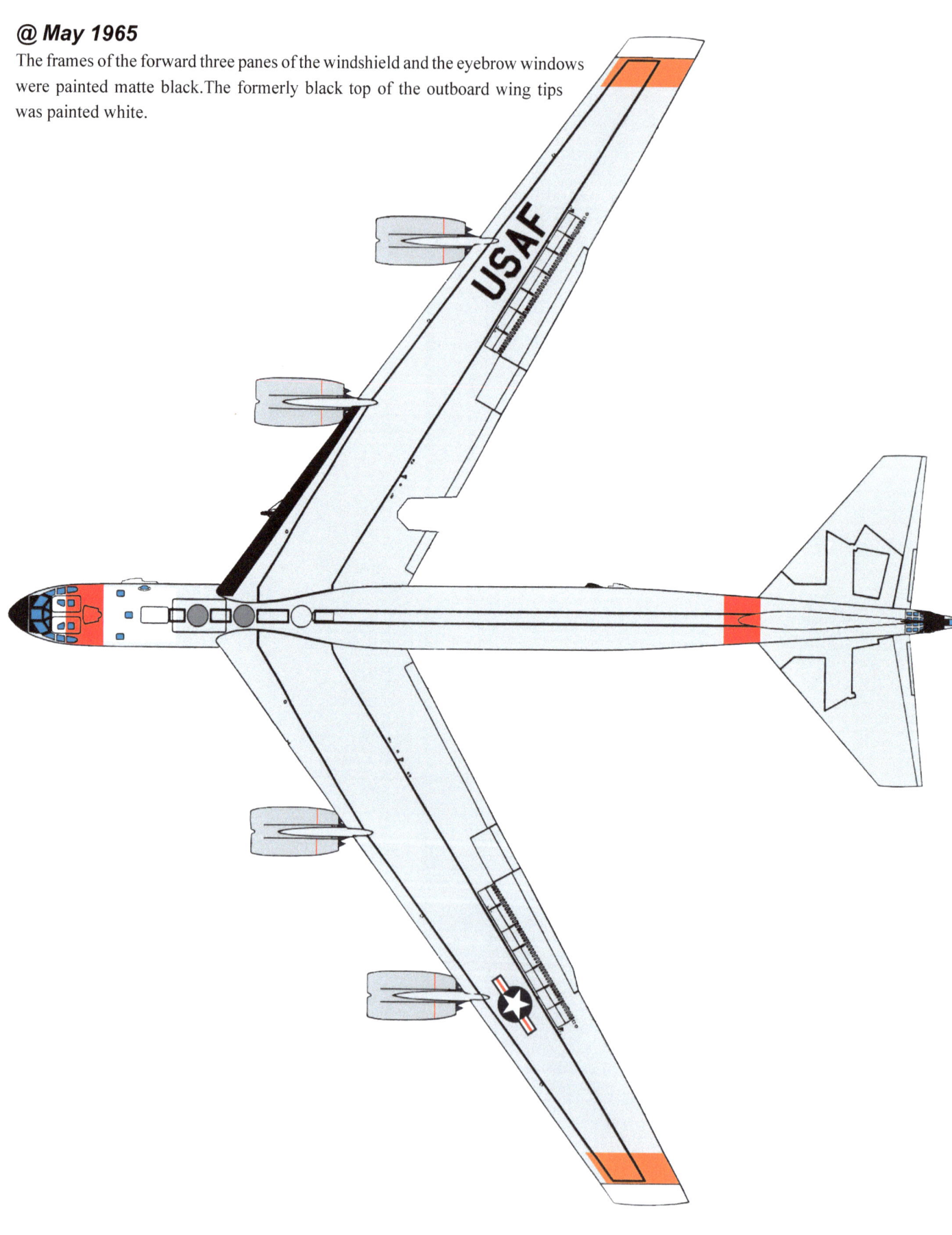

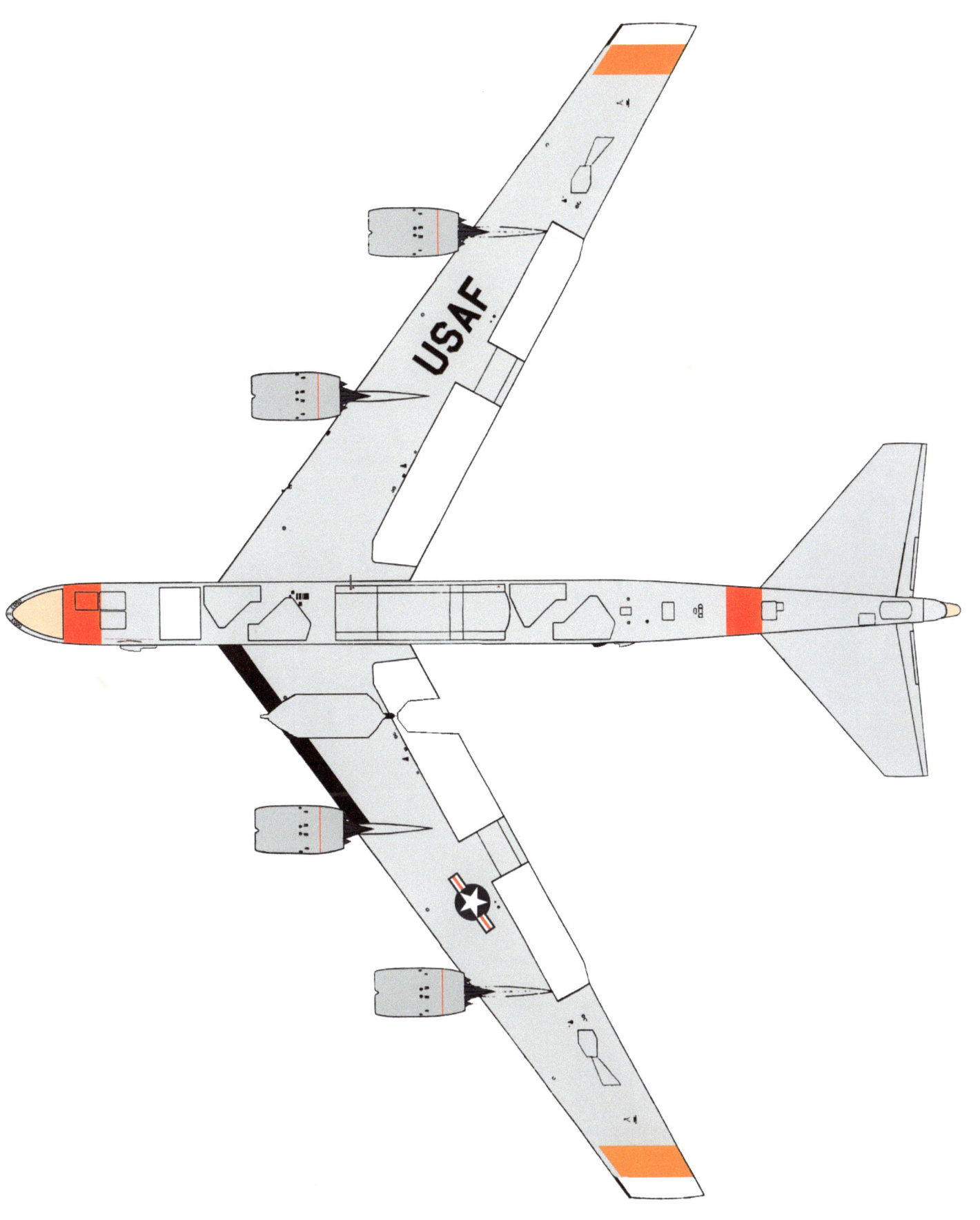

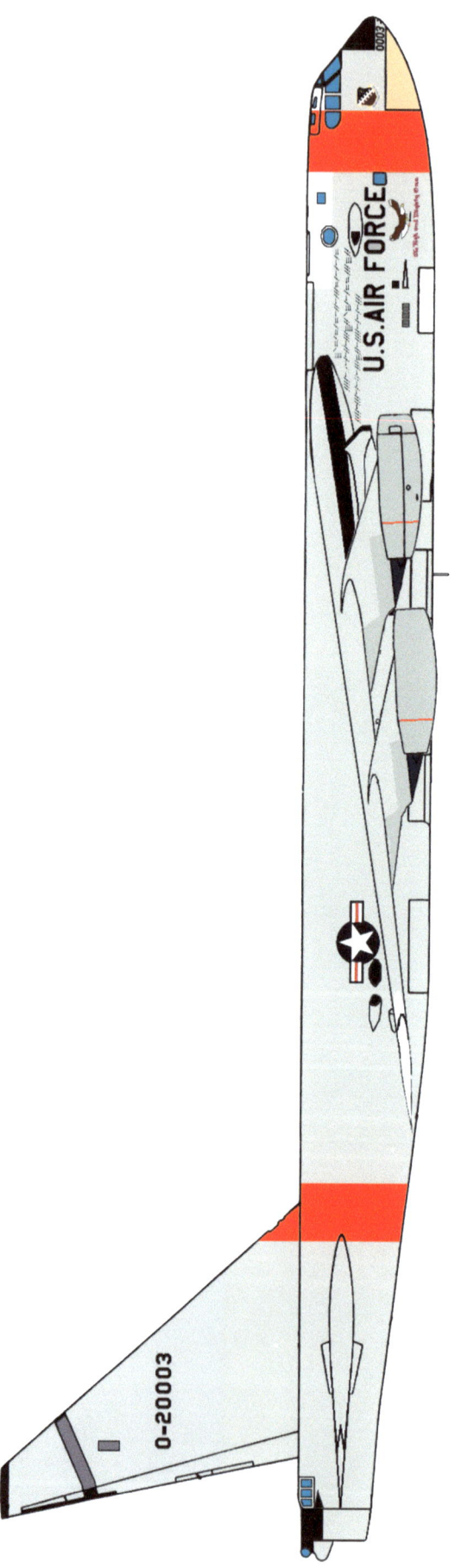
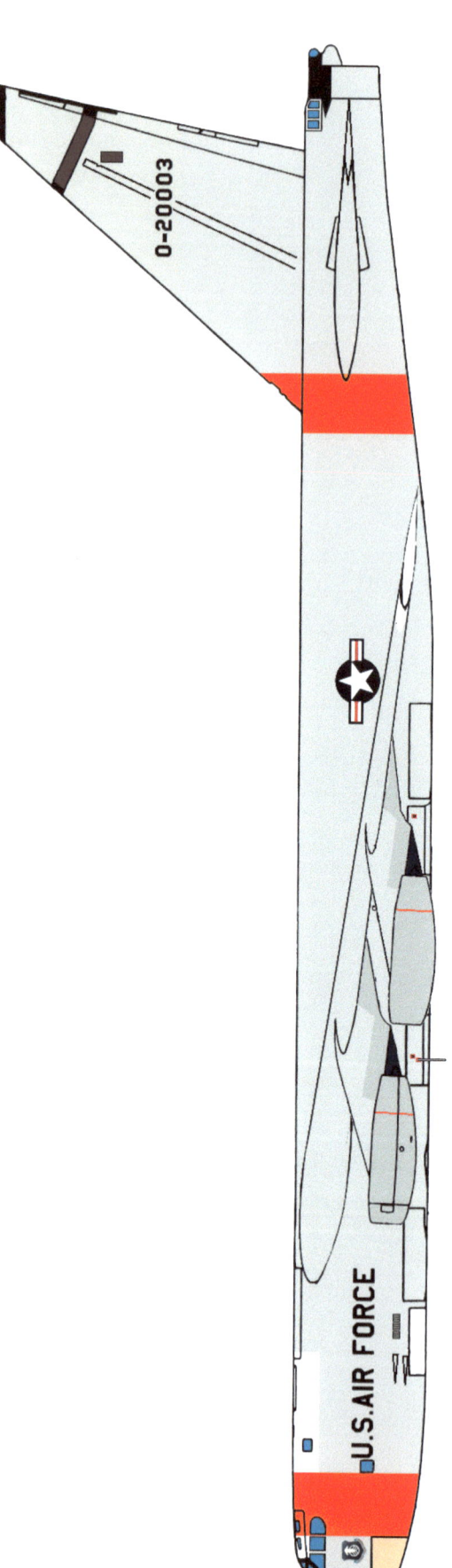

The X-15 mission marks were replaced for the third and last time. A zero was added to the number below the nose radomes so that it read "0003".

The tail turret radome was painted silver.

@ May 1969

The changes described below may have occurred while the NB-52A was at Oklahoma City between November of 1967 and February of 1969.
The fuselage was re-painted silver, removing the last traces of Day-Glo.
The area where the Day-Glo band around the forward fuselage had wrapped over the crew compartment was painted white.

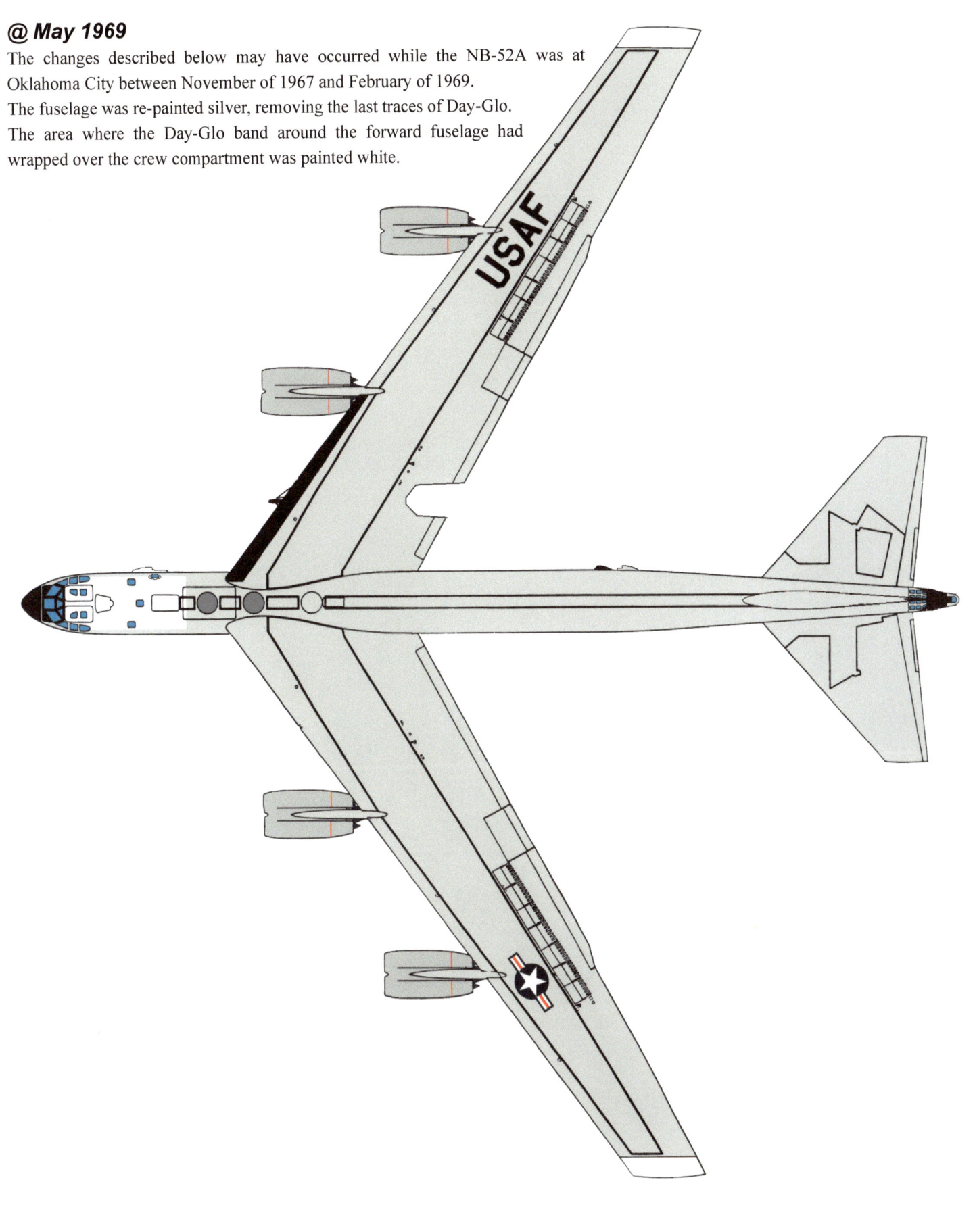

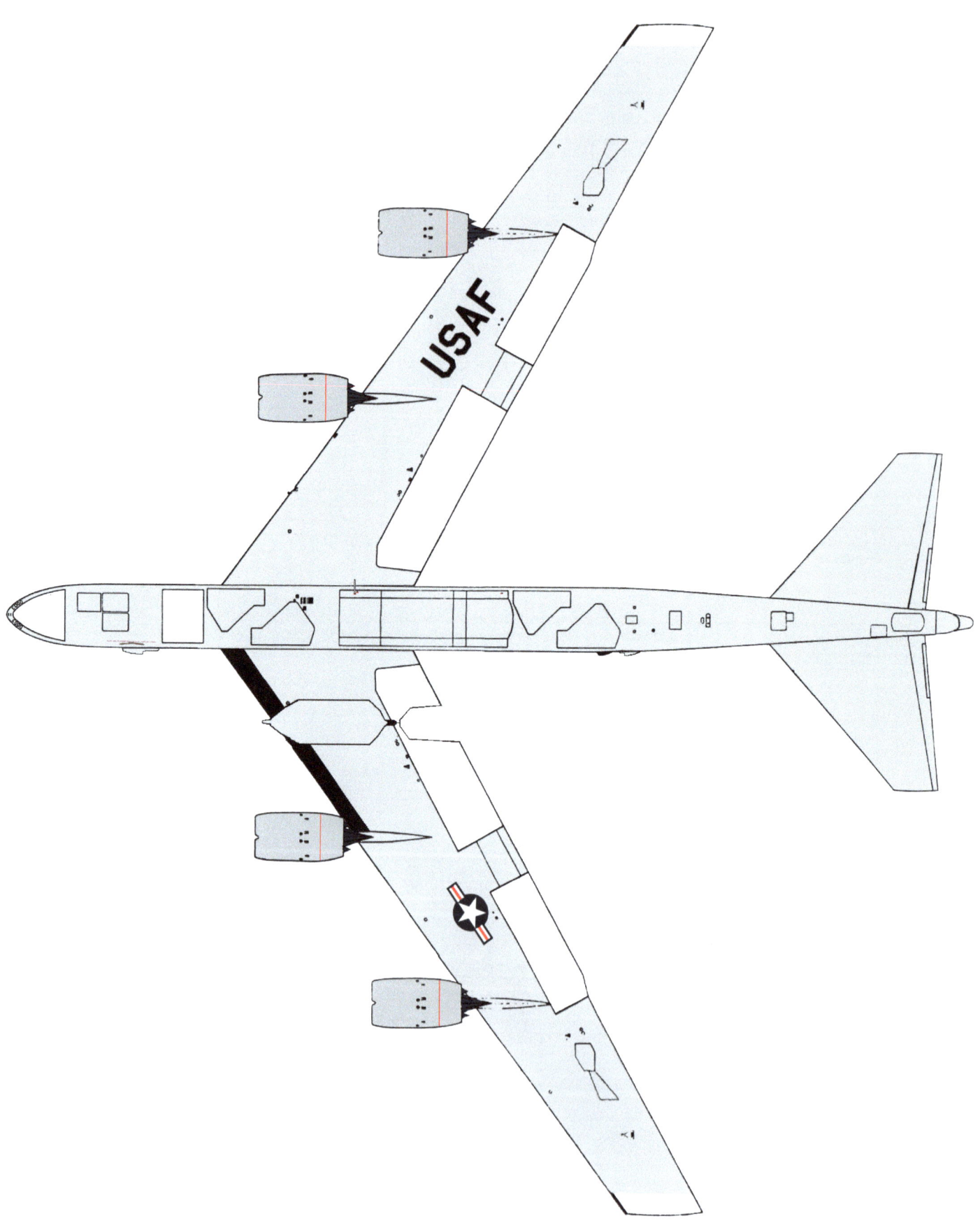

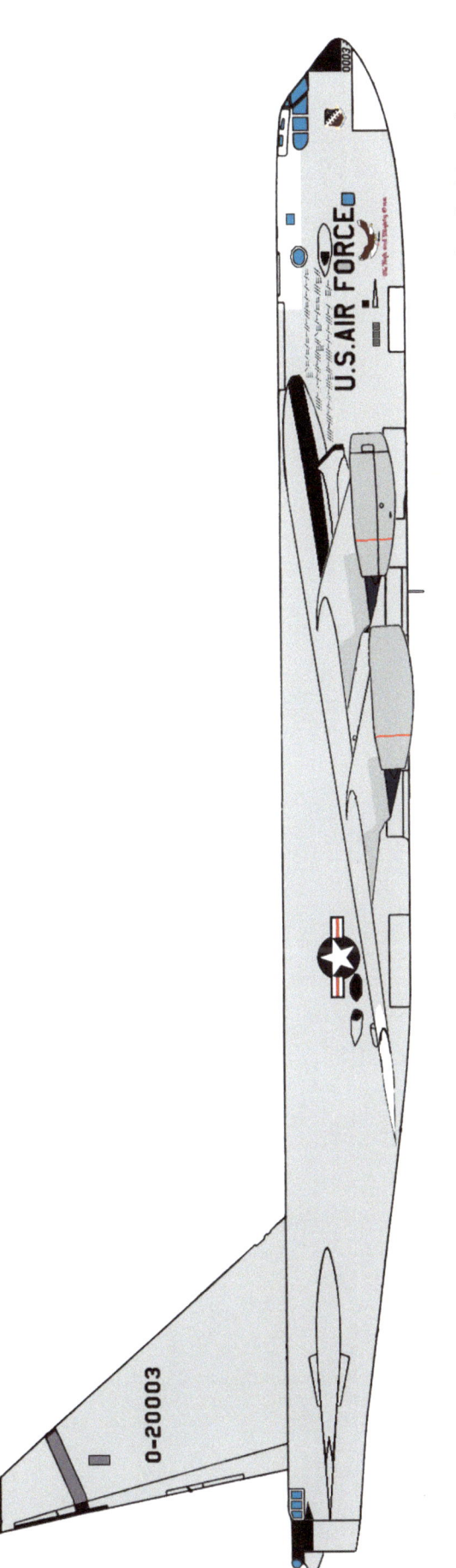
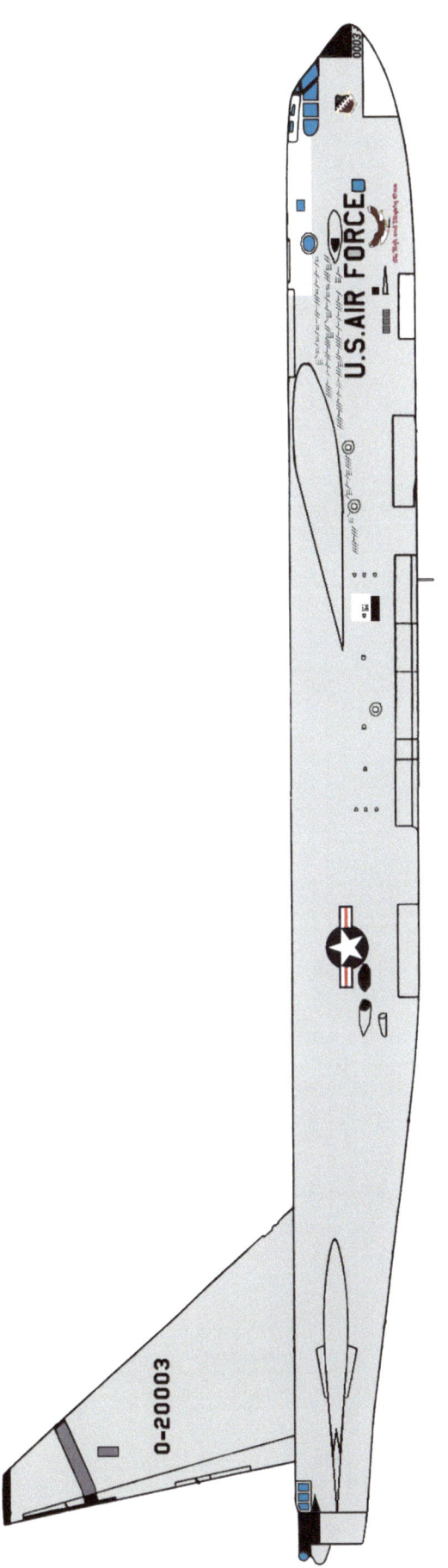

The buff colored chin radome was replaced with a white radome.

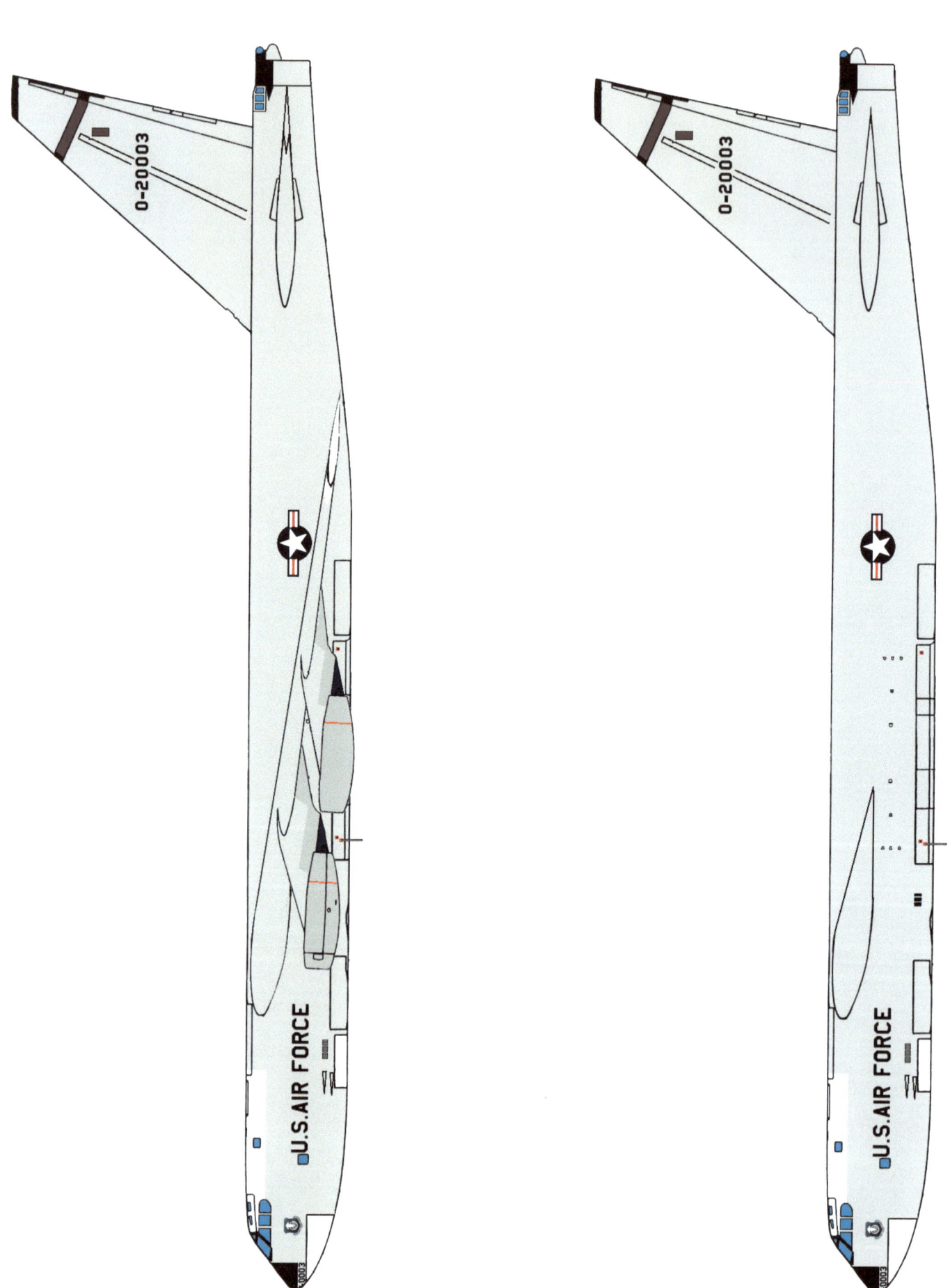

NB-52B 52-0008

@ May 1960

When the NB-52B 52-0008 joined the X-15 program, it was modified in a fashion similar to the NB-52A. There were a few detail differences between the physical configurations of the two airplanes. The tail of the NB-52B was clipped directly aft of the tail gunner's station. There were four louvers above the belly radome rather than the three louvers on the NB-52A. There were two NACA inlets forward of the louvers rather than the single inlet on the NB-52A. The engine pylons had been refitted with slightly shorter versions that had been introduced part way through the production of the B-52D model.

The area above the crew compartment was painted white to reduce the heating effects of sunlight.

There was a fifteen-foot wide band of Day-Glo

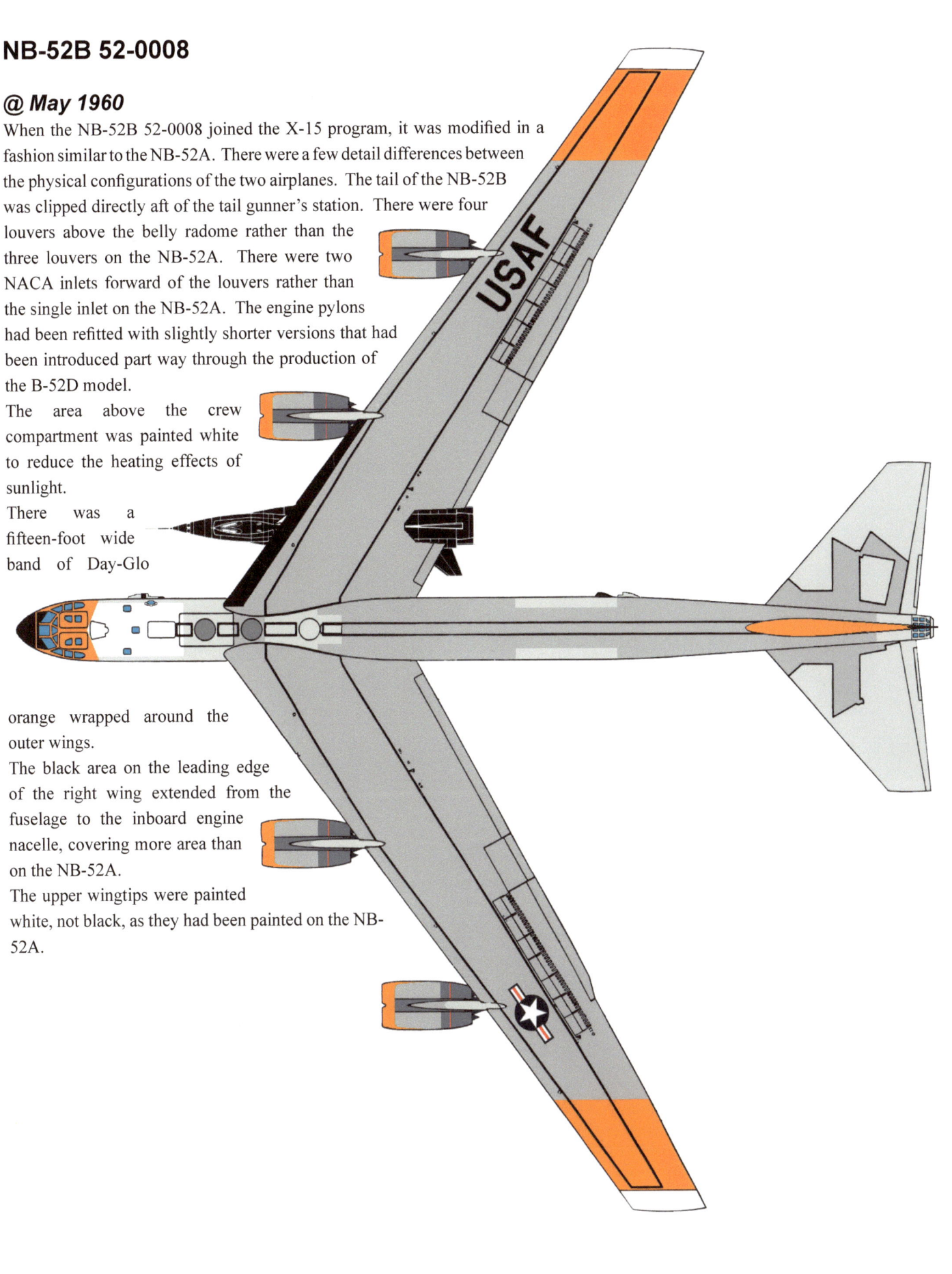

orange wrapped around the outer wings.

The black area on the leading edge of the right wing extended from the fuselage to the inboard engine nacelle, covering more area than on the NB-52A.

The upper wingtips were painted white, not black, as they had been painted on the NB-52A.

The lower part of the bomb bay doors was painted white. The bomb bay doors of the NB-52A had been left unpainted.
The outrigger landing gear doors were painted white as was a rectangular area on the trailing edge of each wing adjacent to the outrigger gear doors.

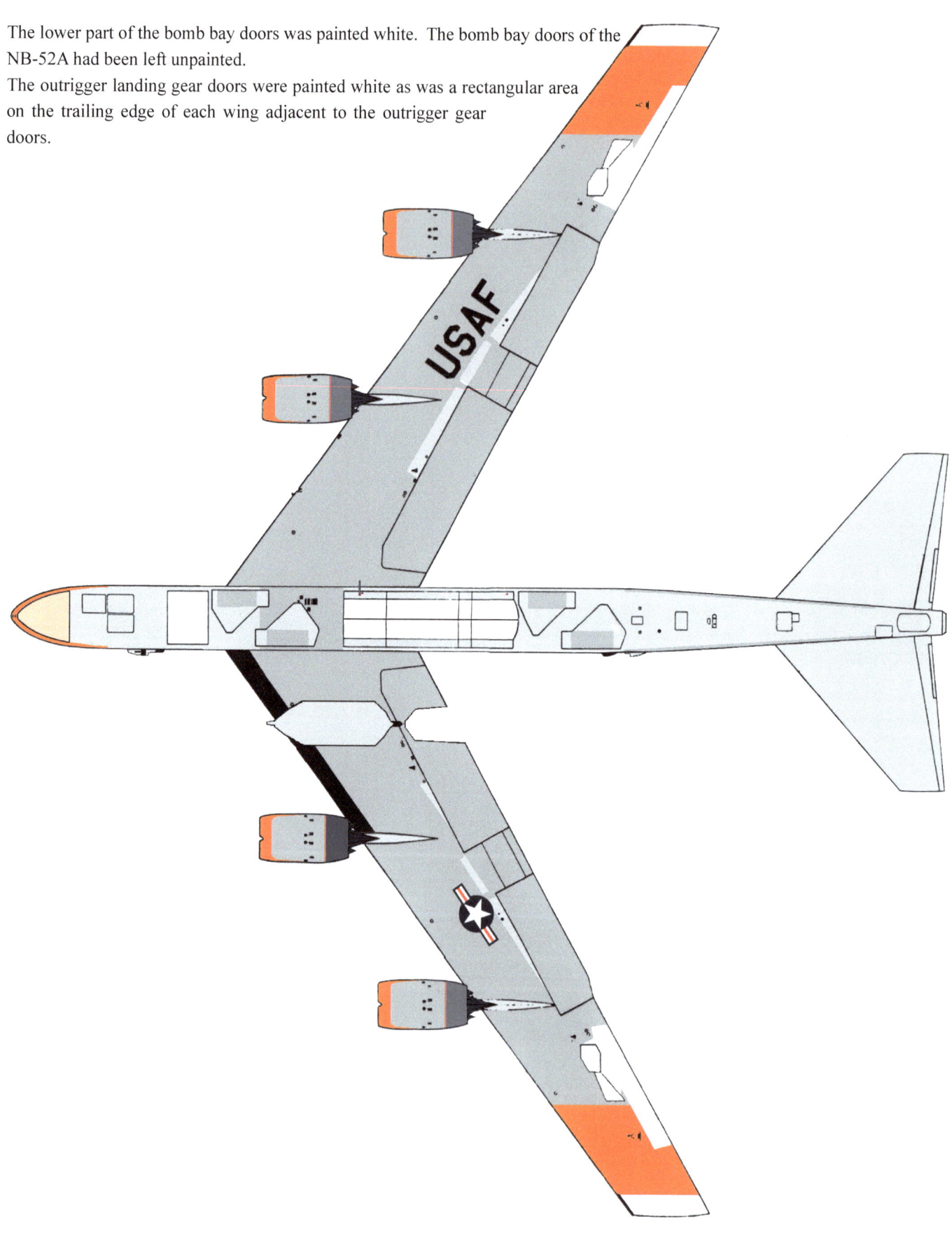

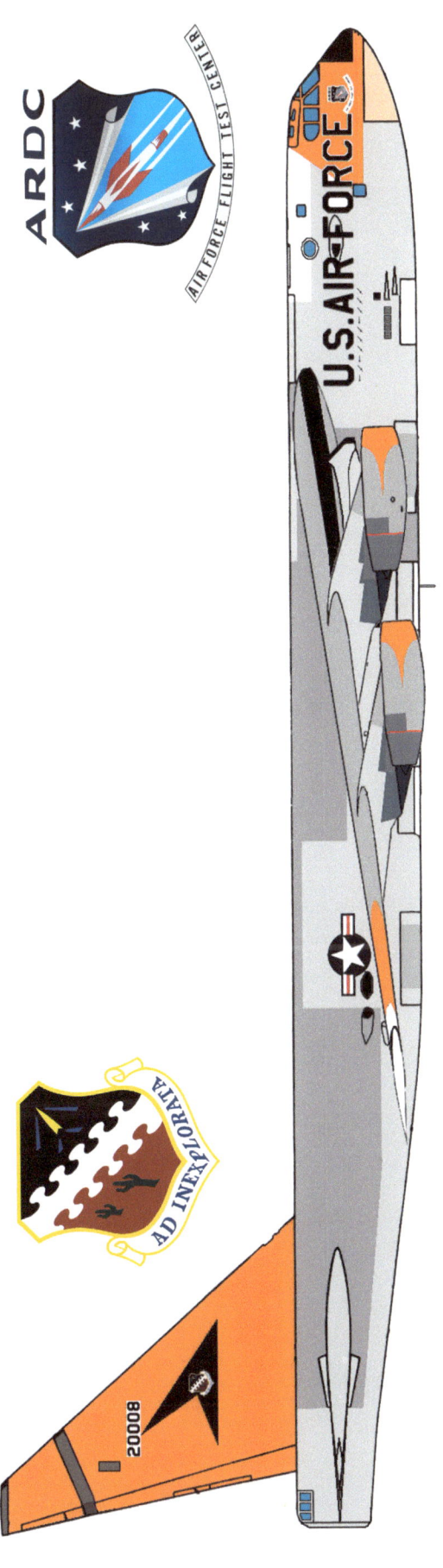
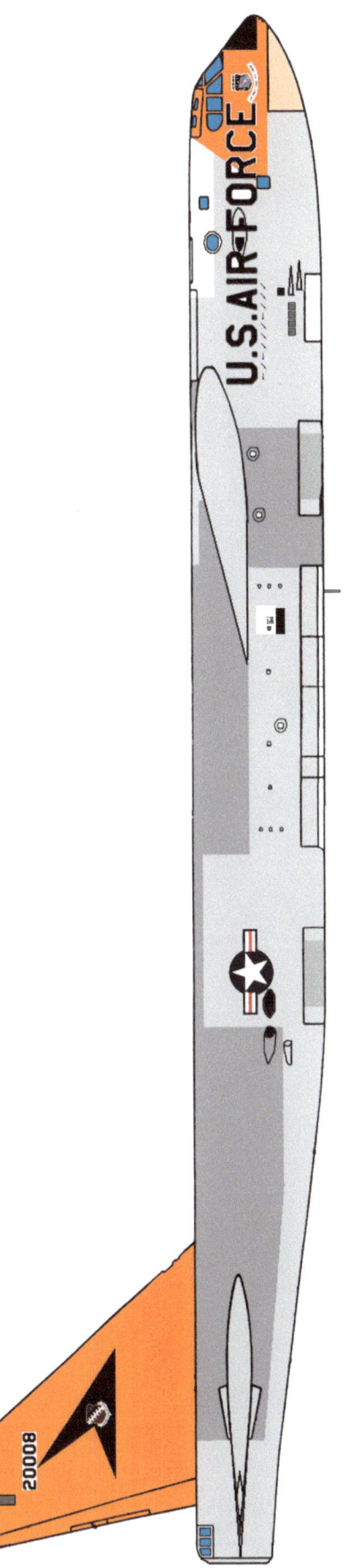

The Day-Glo orange around the nose of 52-0008 extended down to a line level with the top of the chin radome. The trailing edge of the orange area raked forward at a slightly steeper angle than on the NB-52A. It ran from a point near the upper, forward corner of the lower deck window to a point on the top of the fuselage that was even with the aft edge of the rear windows in the pilots' escape hatches.

The Day-Glo around the engine inlets lacked the white cheat line seen on the NB-52A.

The "U. S. AIR FORCE" lettering on the fuselage of the NB-52B was thirty-six inches tall instead of the twenty-four inch tall letters used on the NB-52A. The letters wrapped over the forward video camera housing. The last three letters of FORCE had a bare metal border where they extended into the area painted with Day-Glo orange.

The tail number, "20008", was painted within a bare metal rectangle.

The black chevrons on the vertical stabilizer were slightly different in shape from the ones on the NB-52A. The words "AIR FORCE FLIGHT TEST CENTER" were positioned in an arc ARDC badges were located on either side of the nose below the cockpit. directly below the badge, set on a white background. The acronym ARDC appeared directly above the badge. All of these were set off from the Day-Glo orange by a bare metal outline.

The nose radome was black. The chin radome was buff colored, and the belly radome was white.

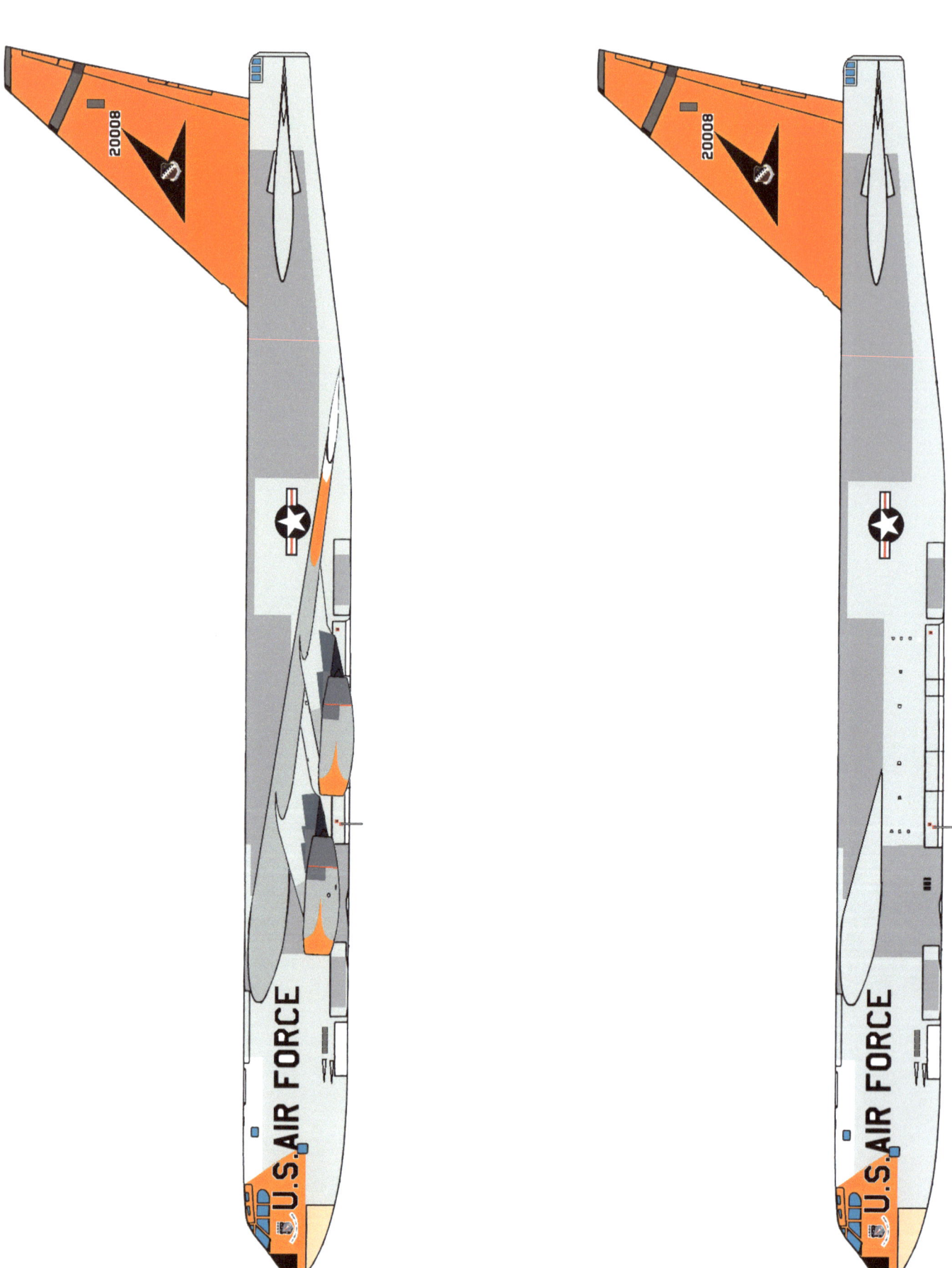

@ September 1960

The NB-52B made an aborted attempt to launch the X-15-1 on June 3, but it did not carry an X-15 until September 2. Sometime during that period of time, the Day-Glo orange was stripped from its vertical stabilizer and a six-foot wide band of Day-Glo orange was painted around the rear fuselage.

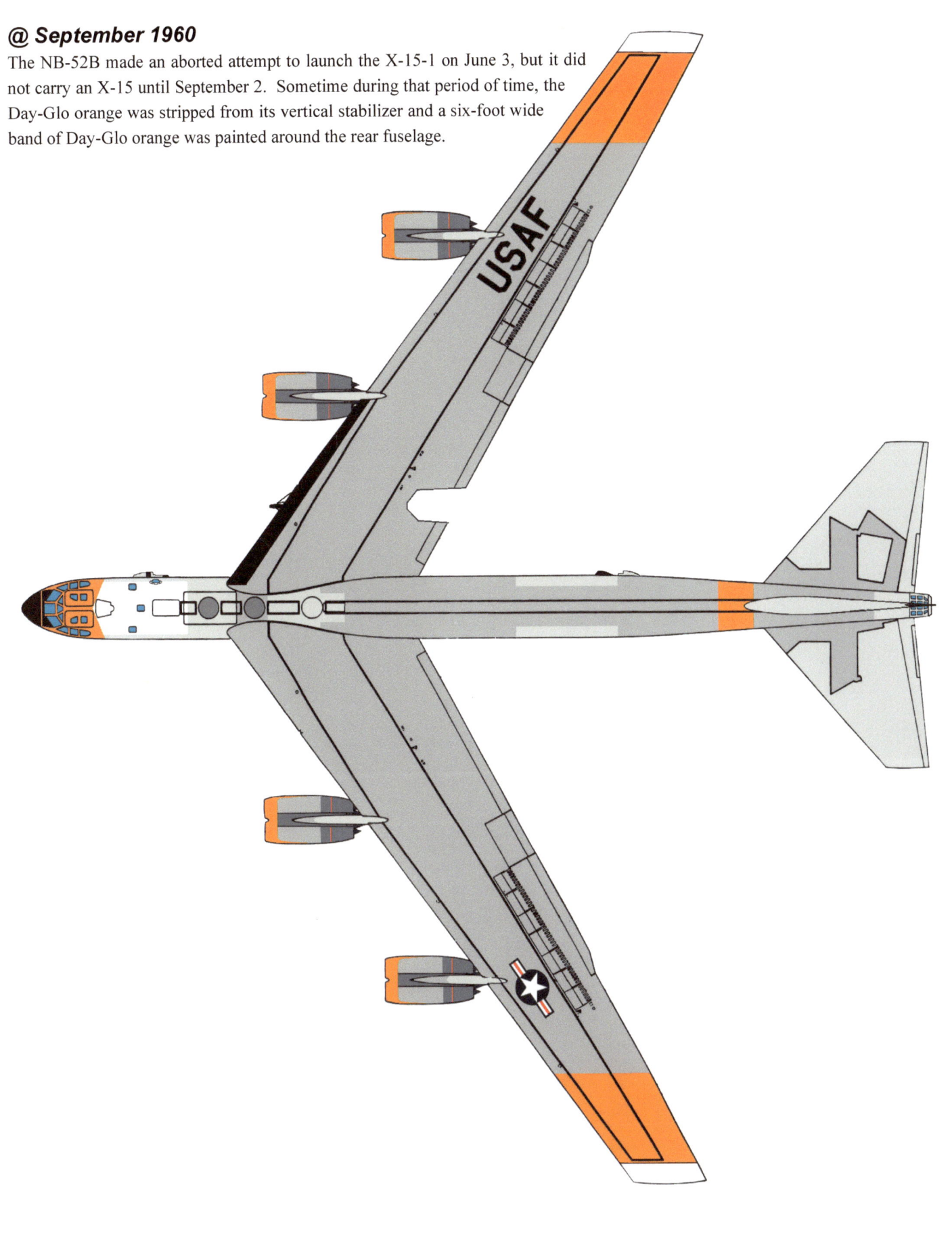

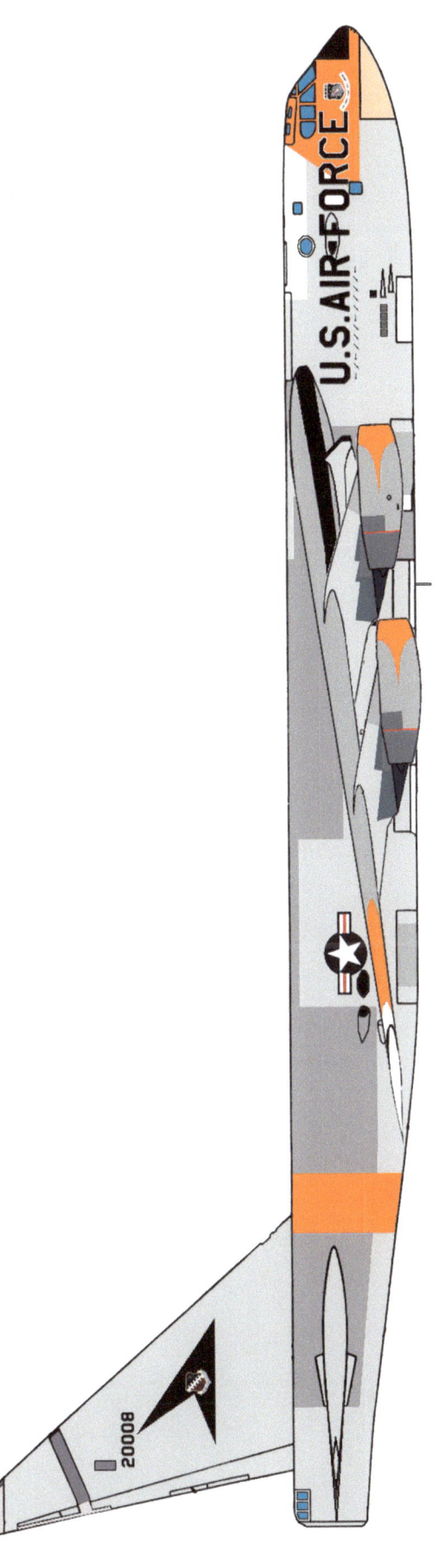
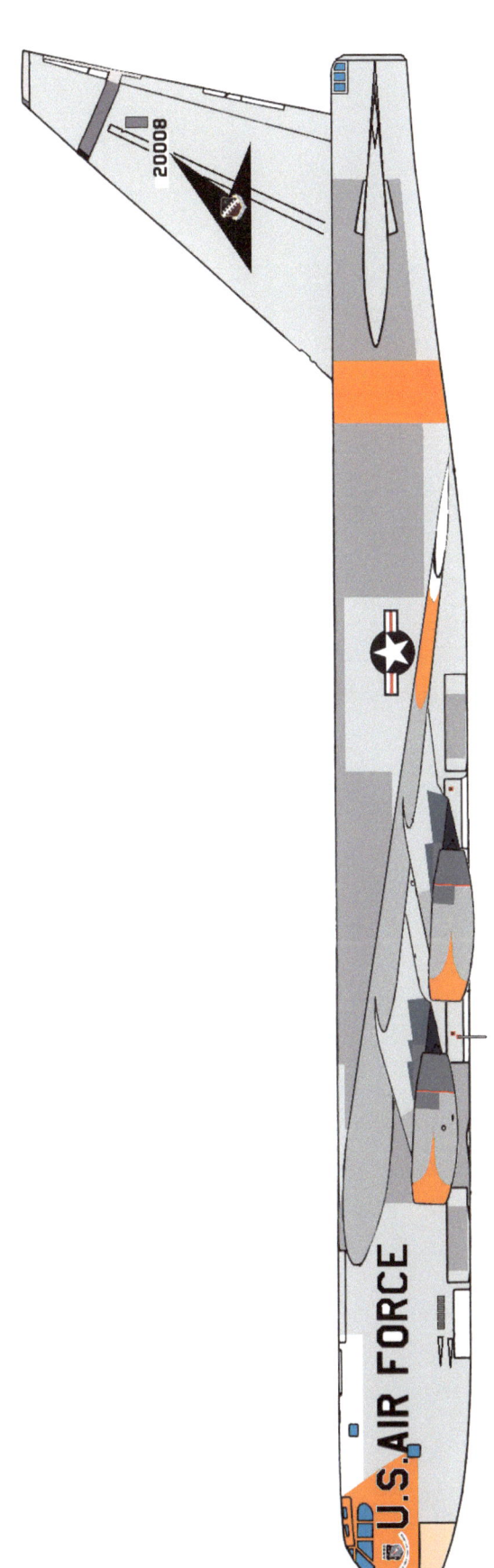

X-15 mission marks were installed below the 36" tall U.S. AIR FORCE on the side of the fuselage.

@ *October 1960*
The Day-Glo orange was stripped from the engine inlets and wingtips.

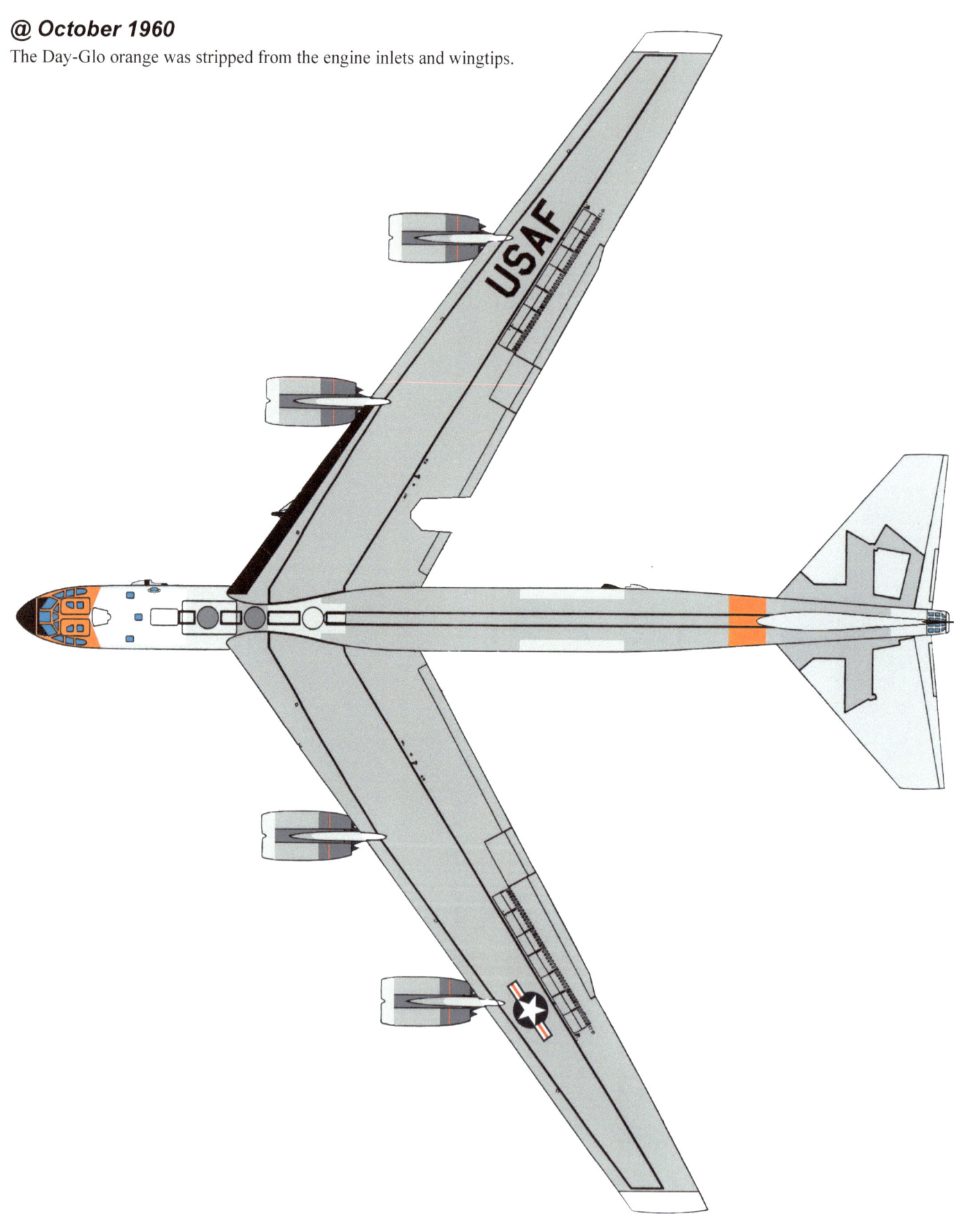

The metal exposed by the removal of the DayGlo orange from the underside of the wing was noticeably brighter than the surfaces that had not previously been painted.

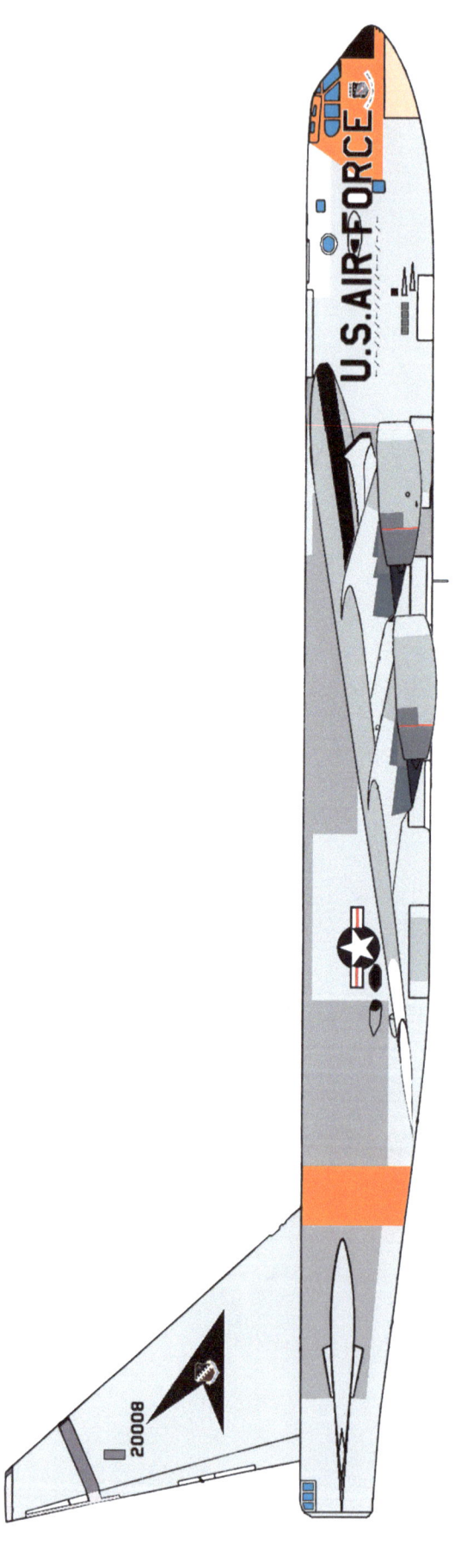

@ *February 1961*

Before February 21, 1961, the NB-52B acquired nose art in the form of an eagle dropping an X-15 set against a blue circle. The eagle carried a ribbon in its beak and the name under the eagle read "The Challenger". The upper part of the blue circle was clipped by the lower edge of the Day-Glo below the cockpit.

@ *May 1961*

Before the Open House on May 21, 1961, the left wingtip was painted black. A six-foot wide band of Day-Glo was painted near each wingtip. This may have been at the same time that the nose art was applied.

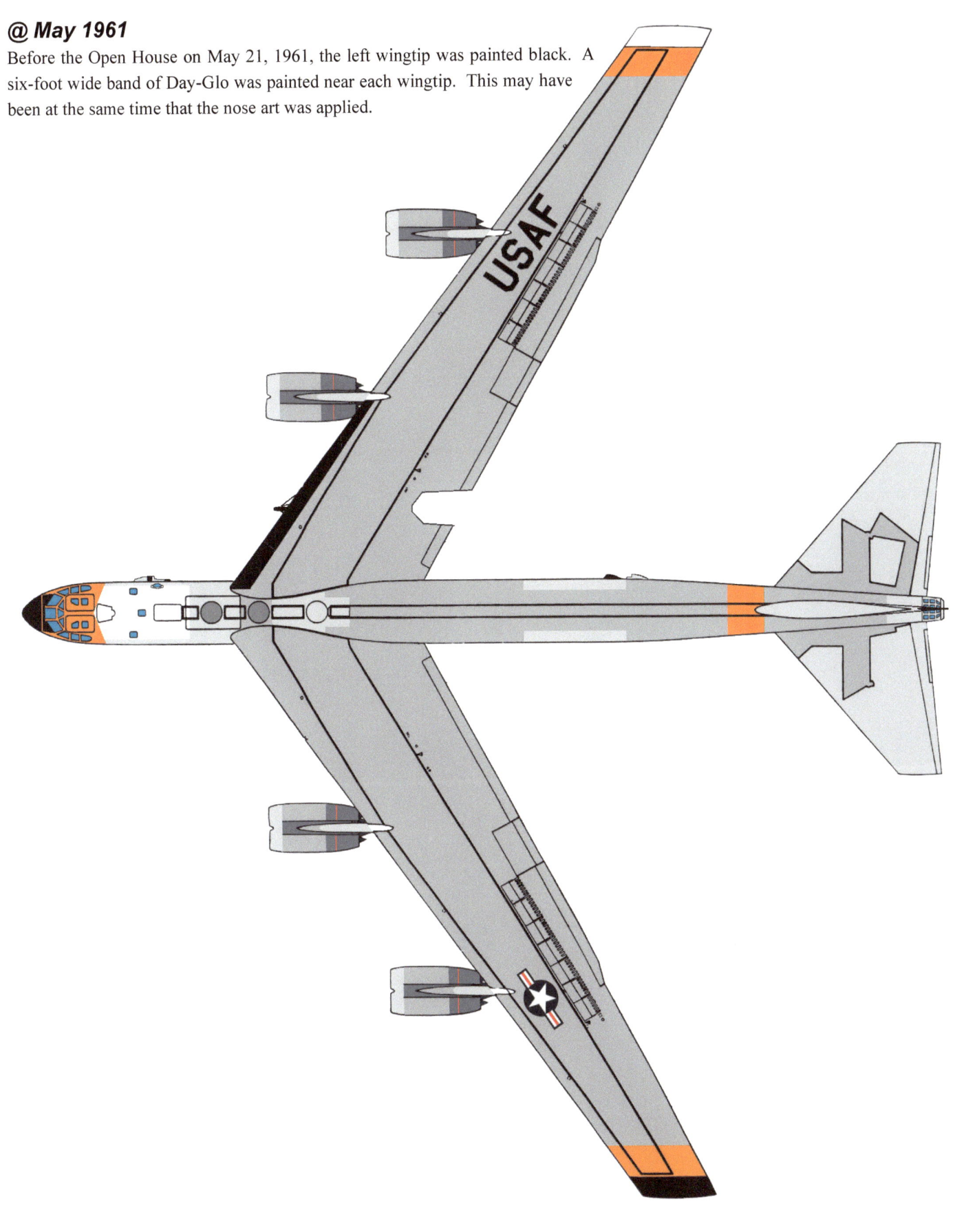

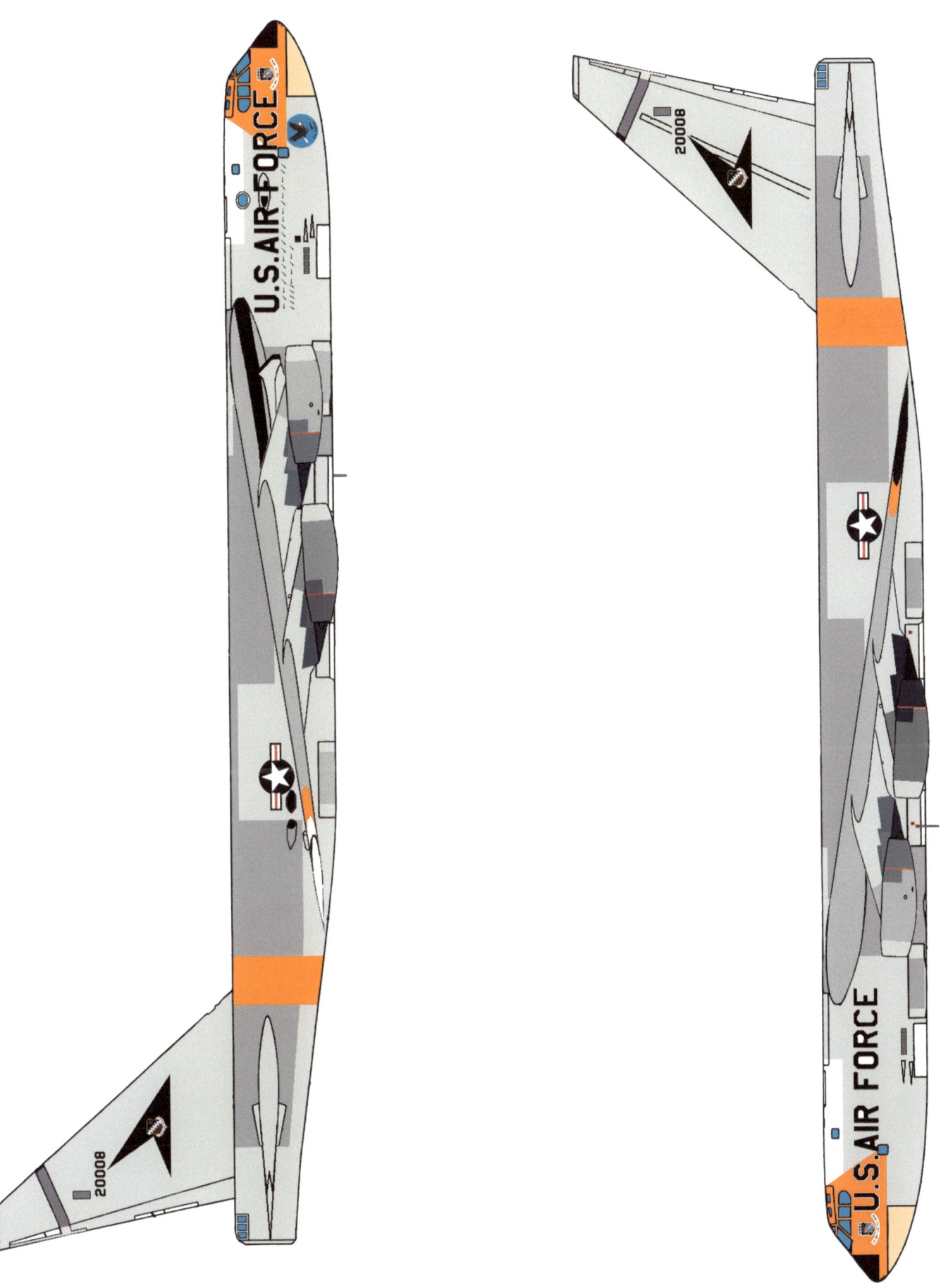

@ *November 1961*

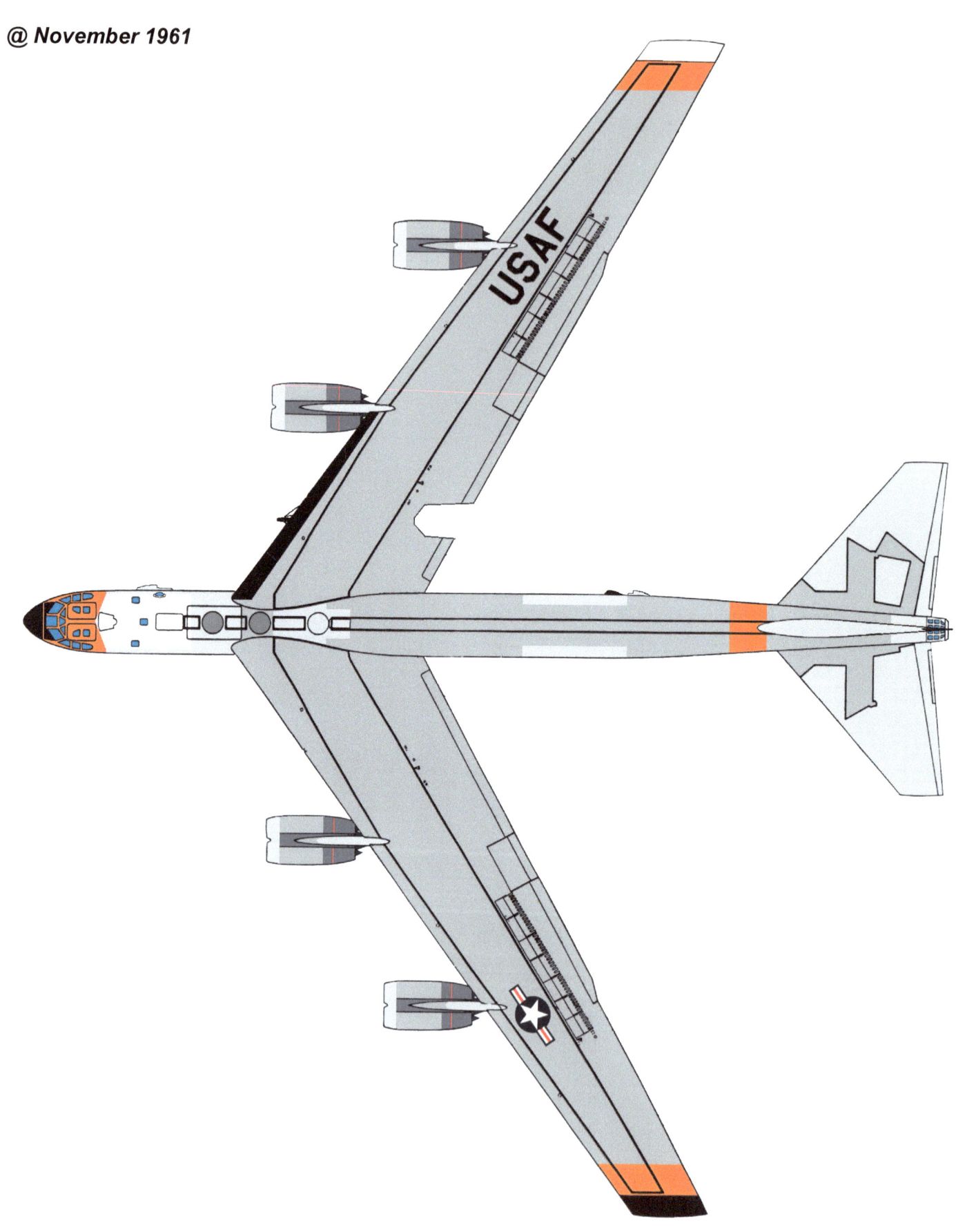

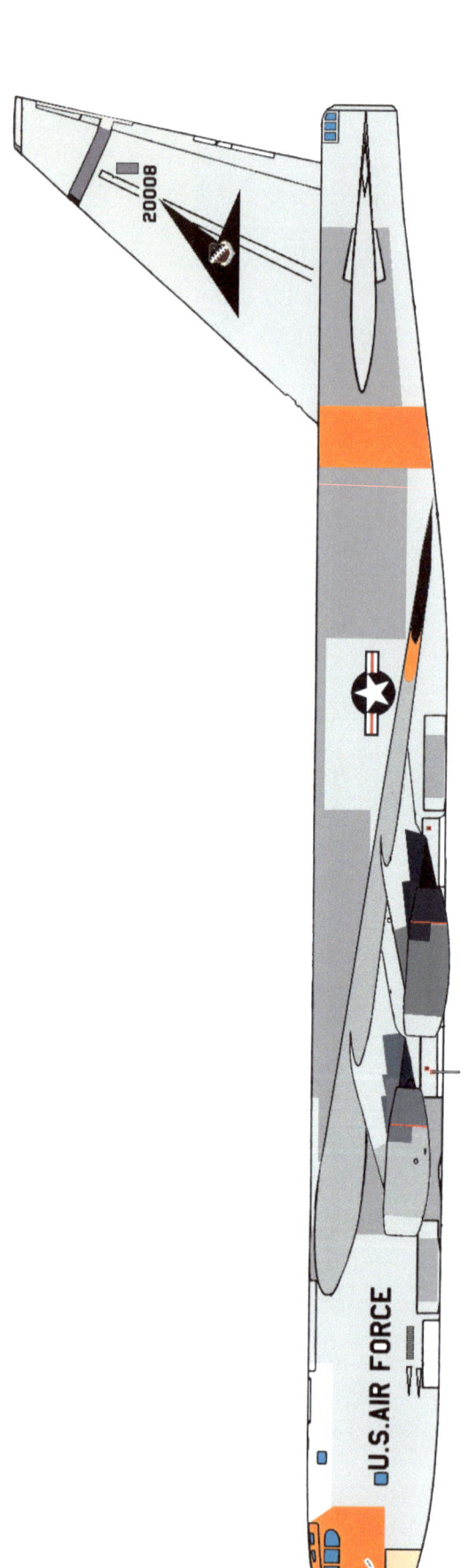

Between the Open House on May 21 and the twenty-first flight of the X-15-2 on November 9, 1961, the thirty-six inch tall "U. S. AIR FORCE" lettering on the fuselage of the NB-52B was changed to smaller twenty-four inch tall letters.

The X-15 mission marks were moved to a location between the leading edge of the wing and the astrodome. A stack of red mission marks on the left end represented flights with an inertial systems pod that simulated the navigation functions of the X-15.

The rear edge of the Day-Glo around the cockpit was stripped forward to a vertical line midway between the chin radome and the lower deck window. The removal of the Day-Glo revealed the left side of the top of the blue circle of the nose art.

The X-15 mission marks, which had been placed below the previous taller letters, were removed and replaced above the new lettering.

@ May 1962

When the NB-52B was sent to Boeing at Wichita, Kansas in early 1962 for modification, the exposed metal areas were coated in silver paint. The left wing tip was painted white.

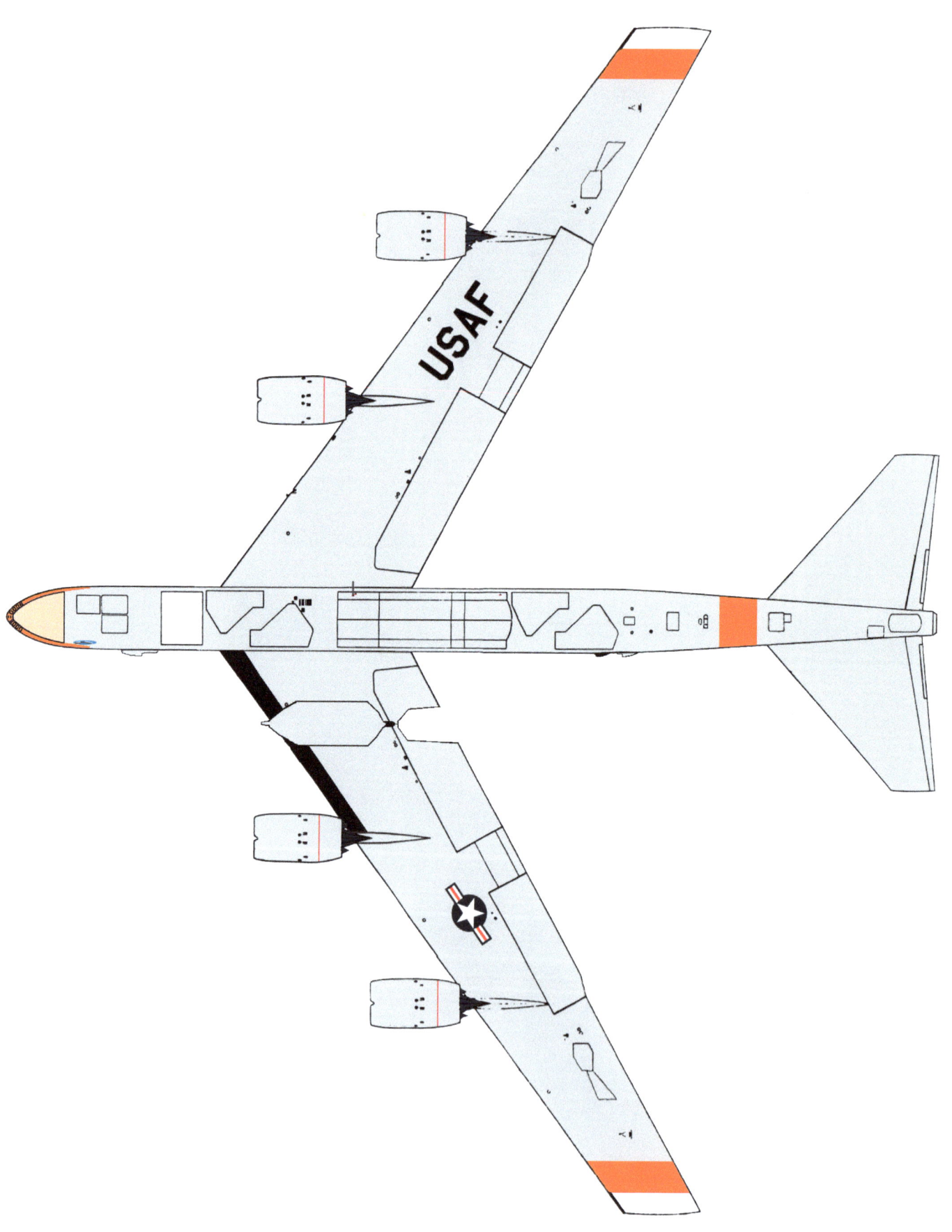

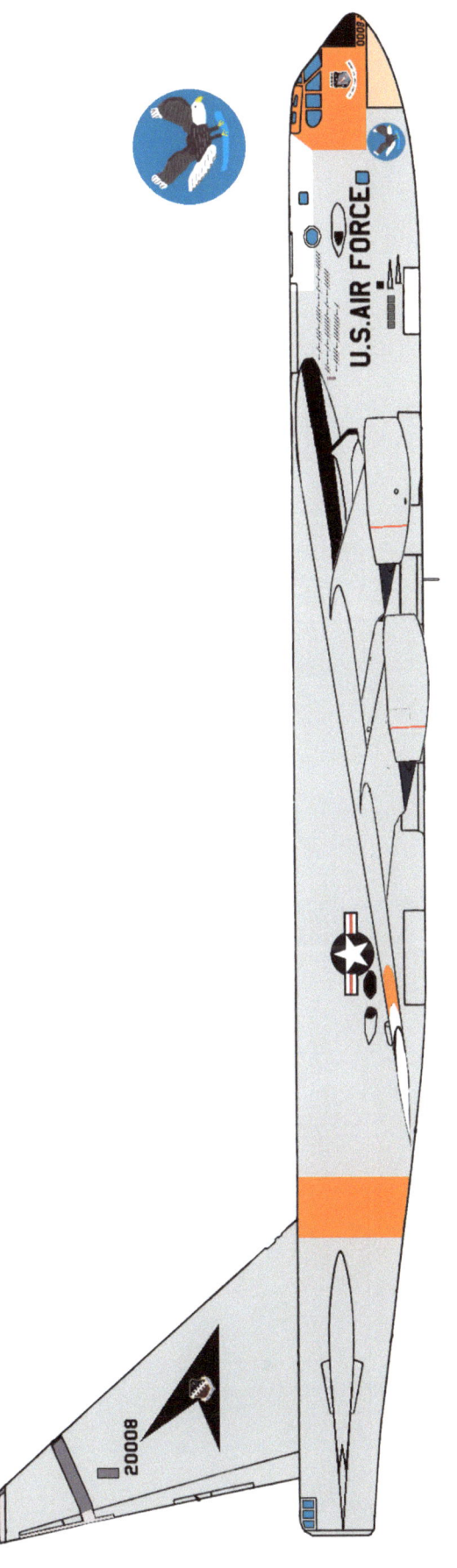
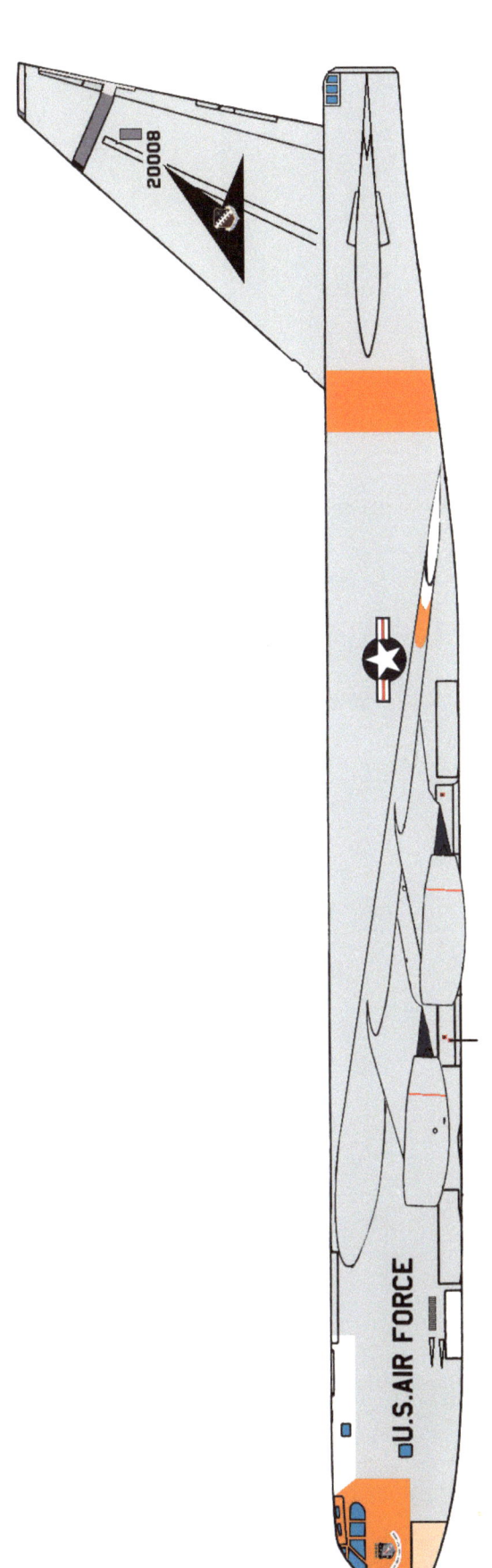

The X-15 mission marks were replaced a second time with a more compact set. The last four digits of the Air Force serial number, 0008, were added on each side of the nose below the nose radome. The eagle nose art was painted over. A slightly different eagle on a blue circle was painted lower on the nose. The blue circle was no longer truncated on the top at the edge of the area painted Day-Glo orange.

@ *May 1963*

A Day-Glo red band was painted around the forward fuselage between the chin radome and the lower deck window. The area around the cockpit forward of the Day-Glo band was painted silver. This included the area above the cockpit

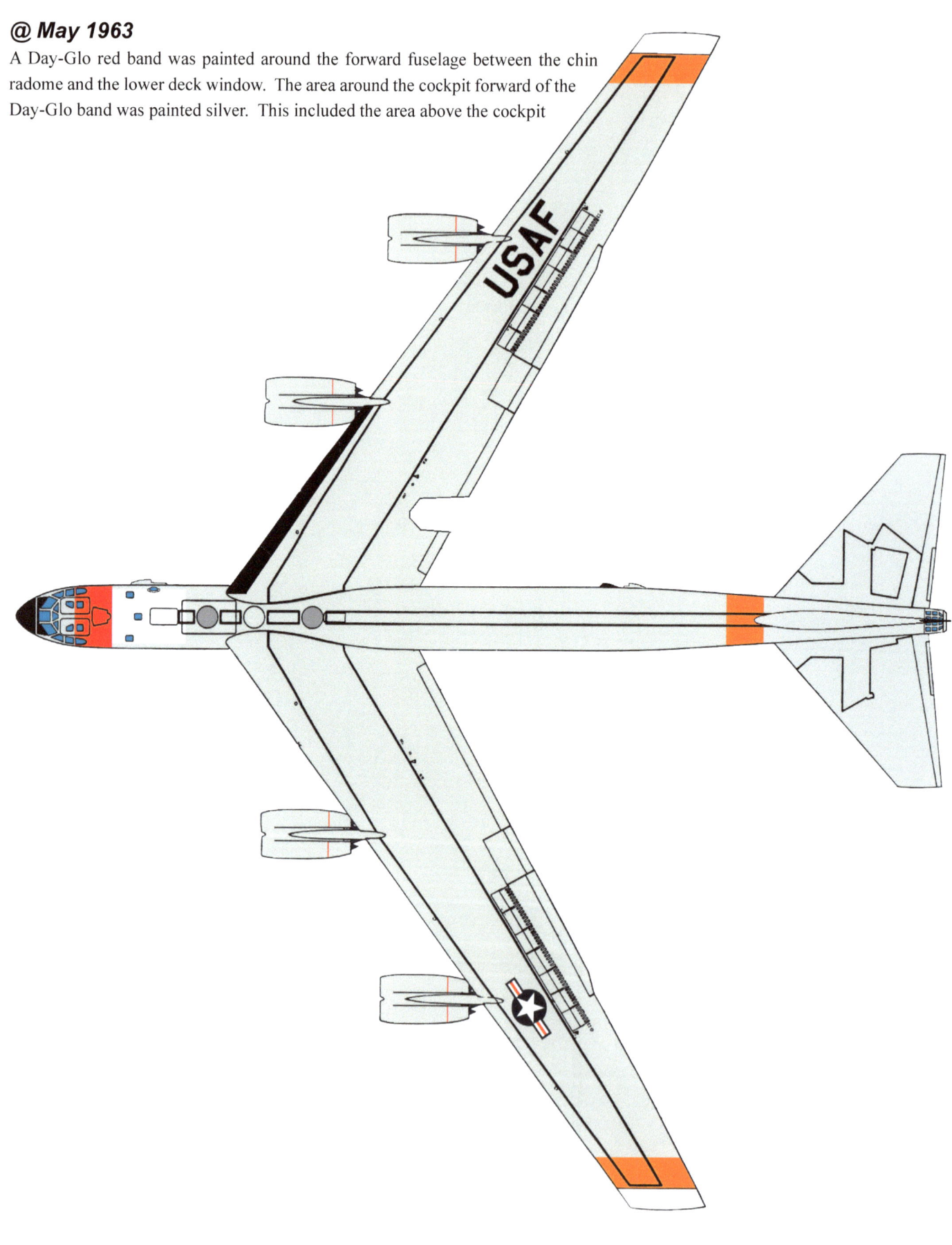

The white belly radome was replaced with a buff colored radome.

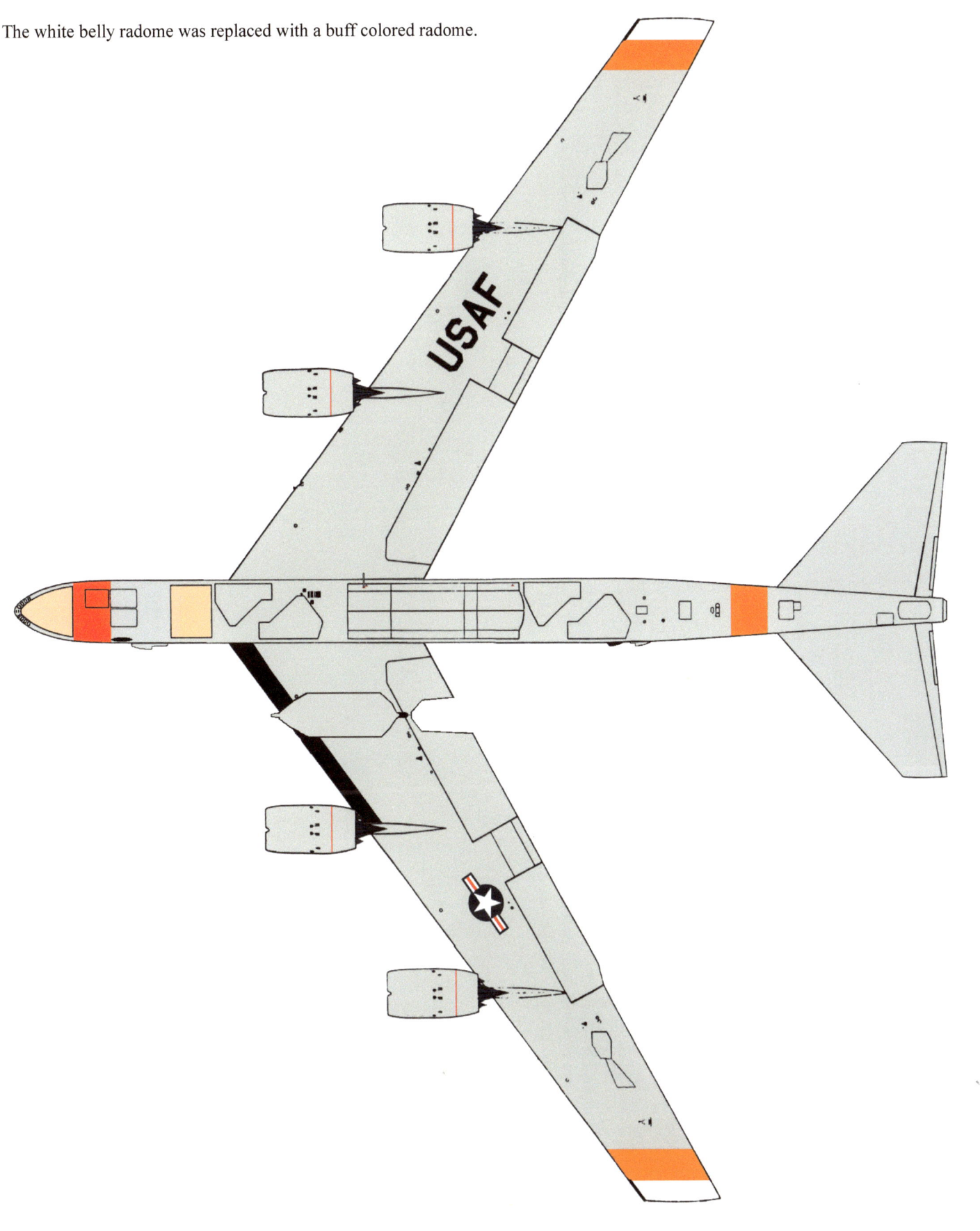

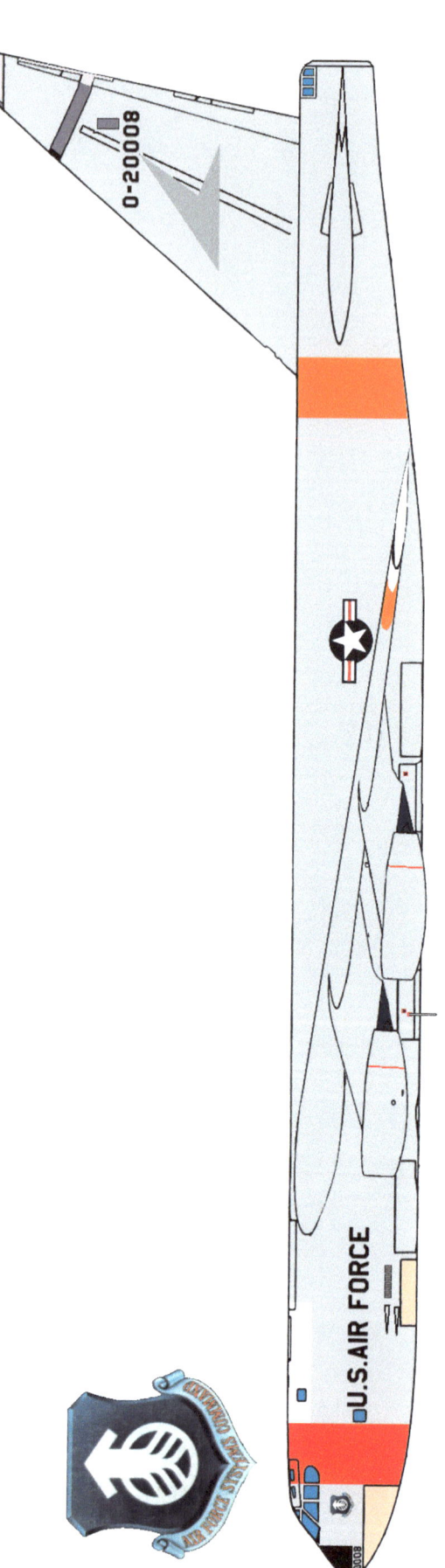

A black circle was painted below the lower deck window. It may have been the location for nose art to be added later. The black tail chevron, which had been applied to airplanes assigned to the AFFTC, was removed. Its removal left a chevron of a slightly different shade of silver on the tail. A leading zero was added to the tail number to reflect the fact that the airplane was over ten years old.

@ *July 1963*
The band of Day-Glo around the forward fuselage was stripped from the underside of the fuselage.

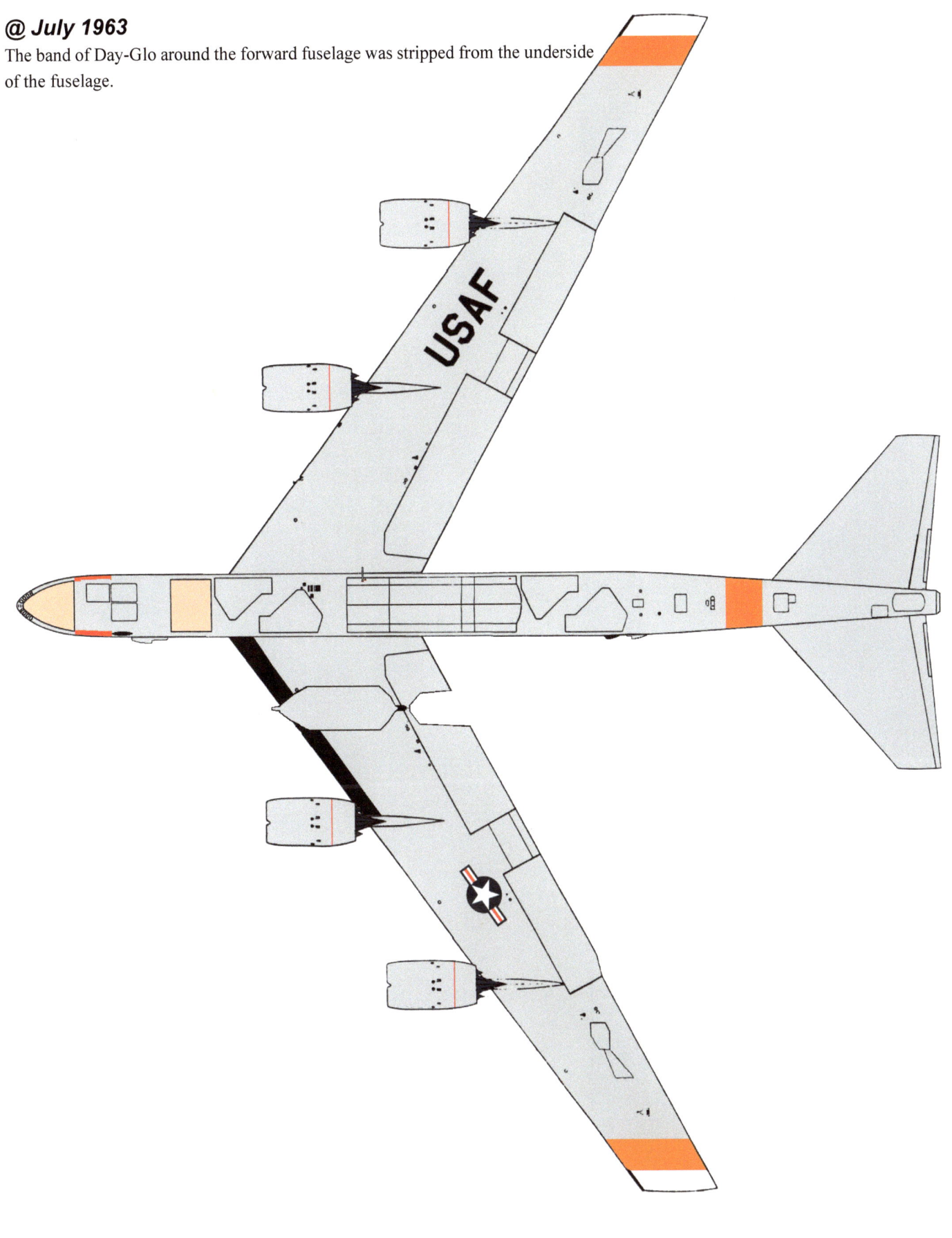

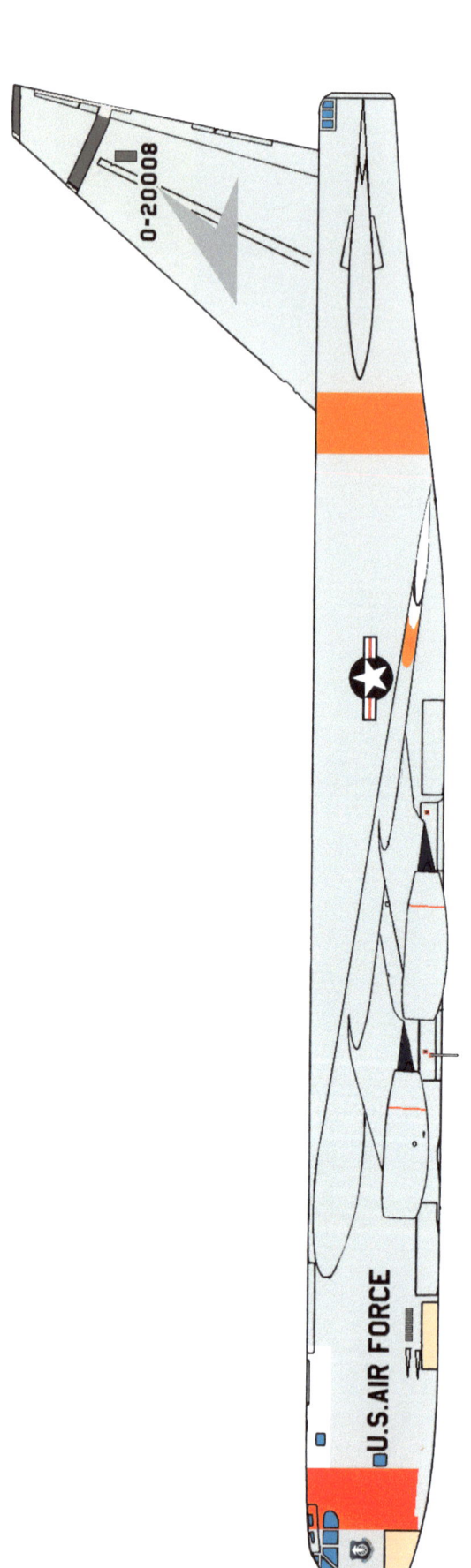

The lower edge of the Dayglo on the right side was a horizontal line. The lower edge on the left side ran along the outboard edge of the access hatch on the bottom of the fuselage.

@ October 1965

All the Day-Glo was removed from the NB-52B.
The white area over the crew compartment was extended forward to a vertical line that was even with the center of the frame between the last two panes of the windshield. The area directly above the pilots remained silver.
The engine nacelles and the aft portions of the engine pylons were painted gloss light gray.

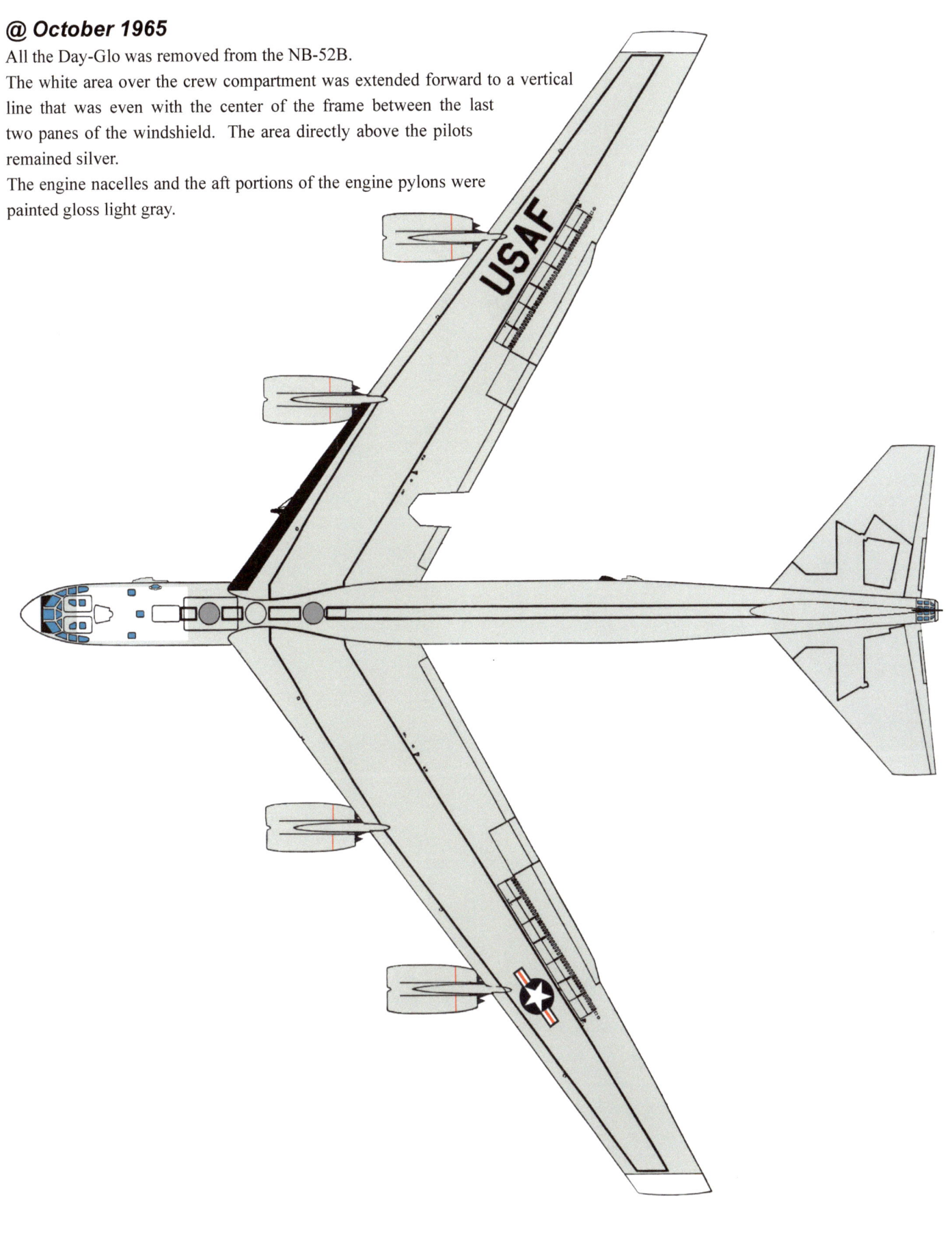

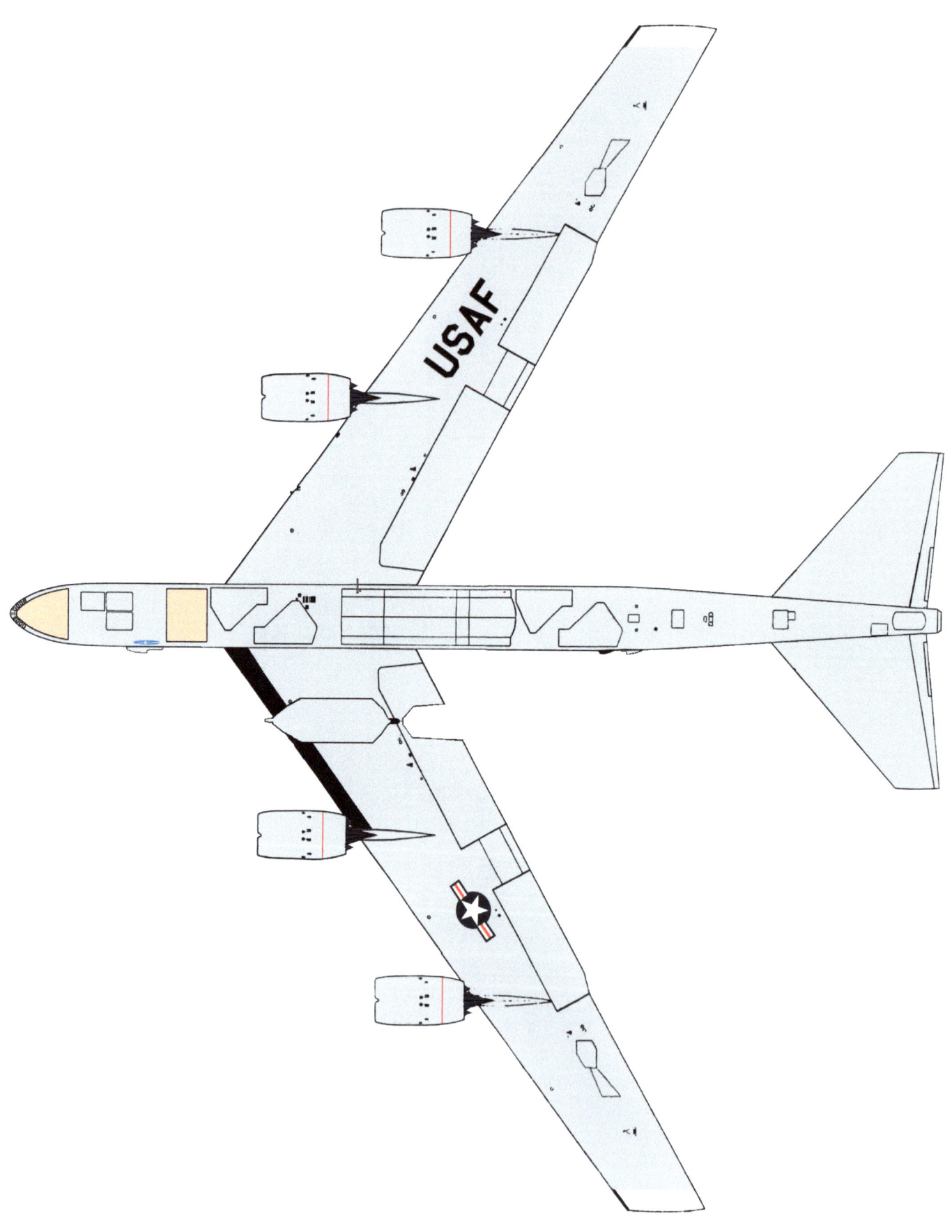

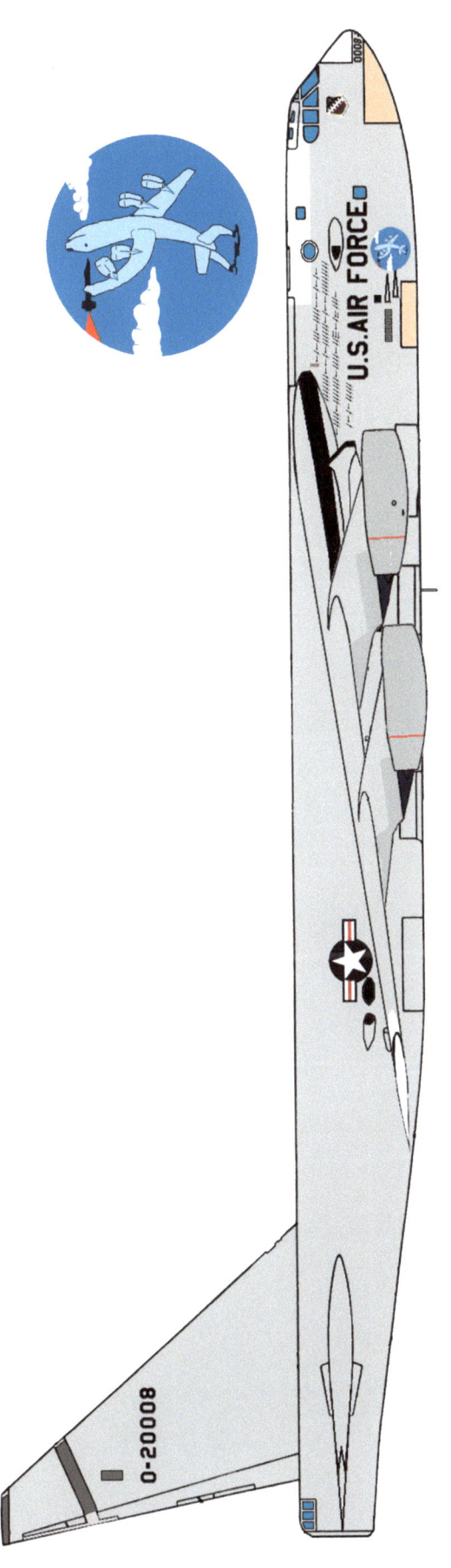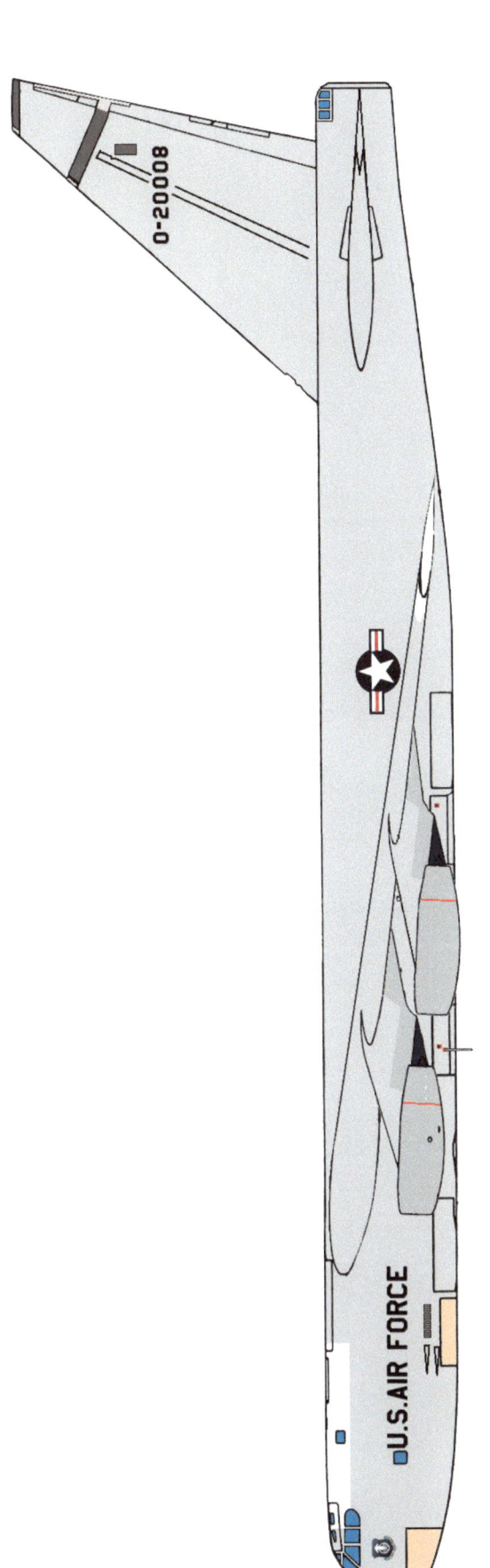

The eagle nose art was replaced by a charicature of a B-52 throwing an X-15 set against a blue circle with white clouds. The X-15 mission marks were revised for the third time. The black upper nose radome of the NB-52B was temporarily replaced with a white radome. The tail number, "0-20008", was moved down and forward on the vertical stabilizer.

@ May 1966
The white nose radome was replaced with a black radome.

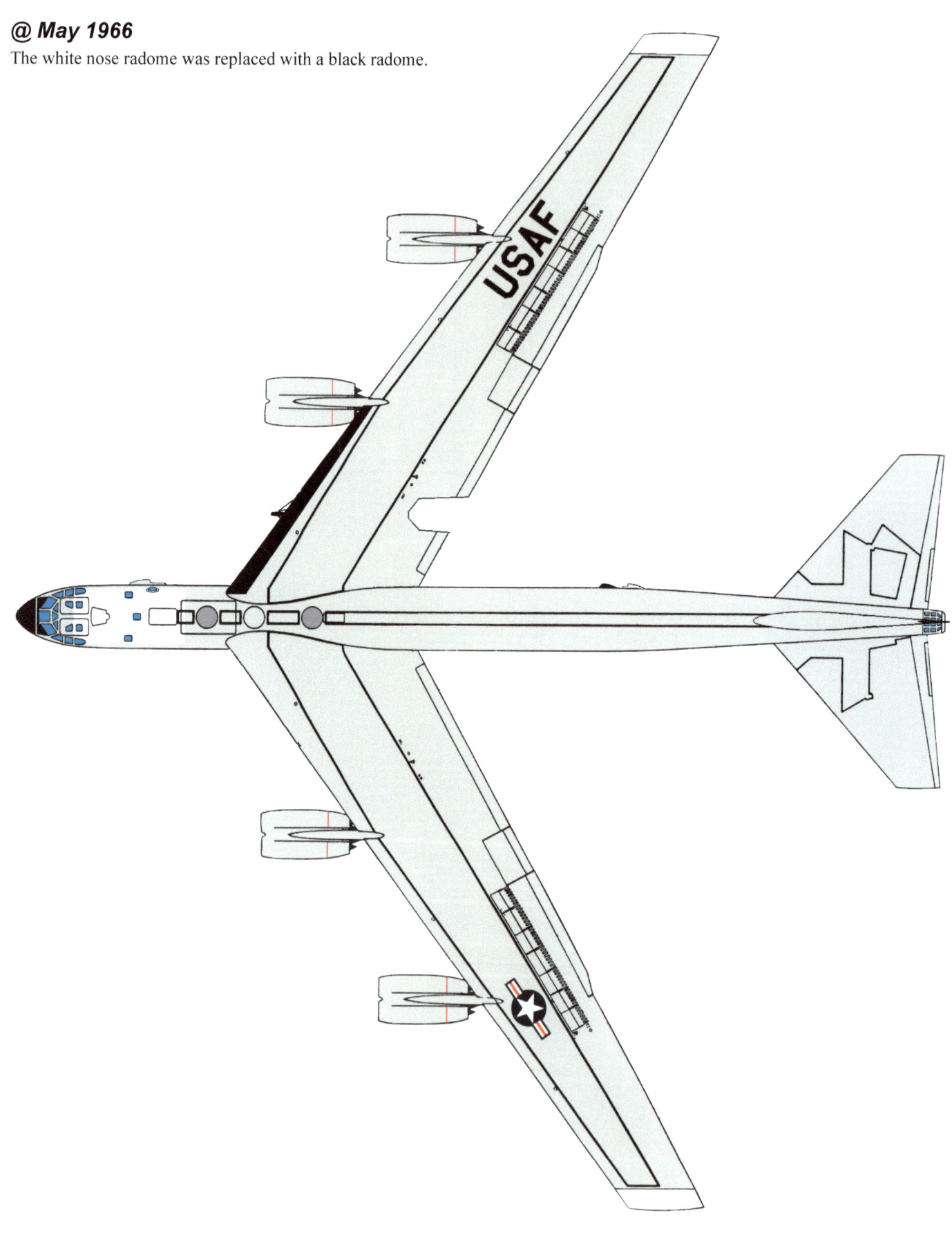

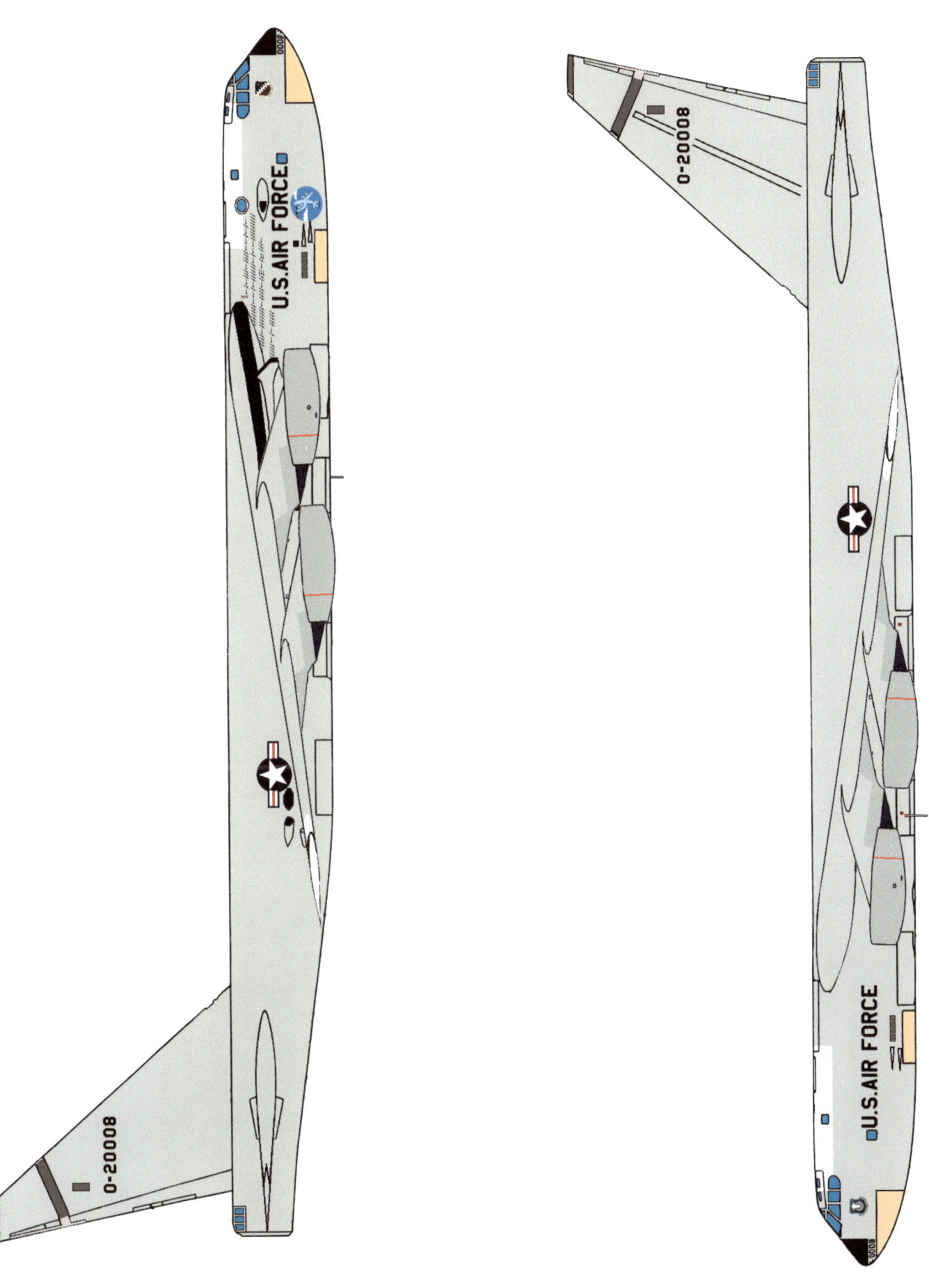

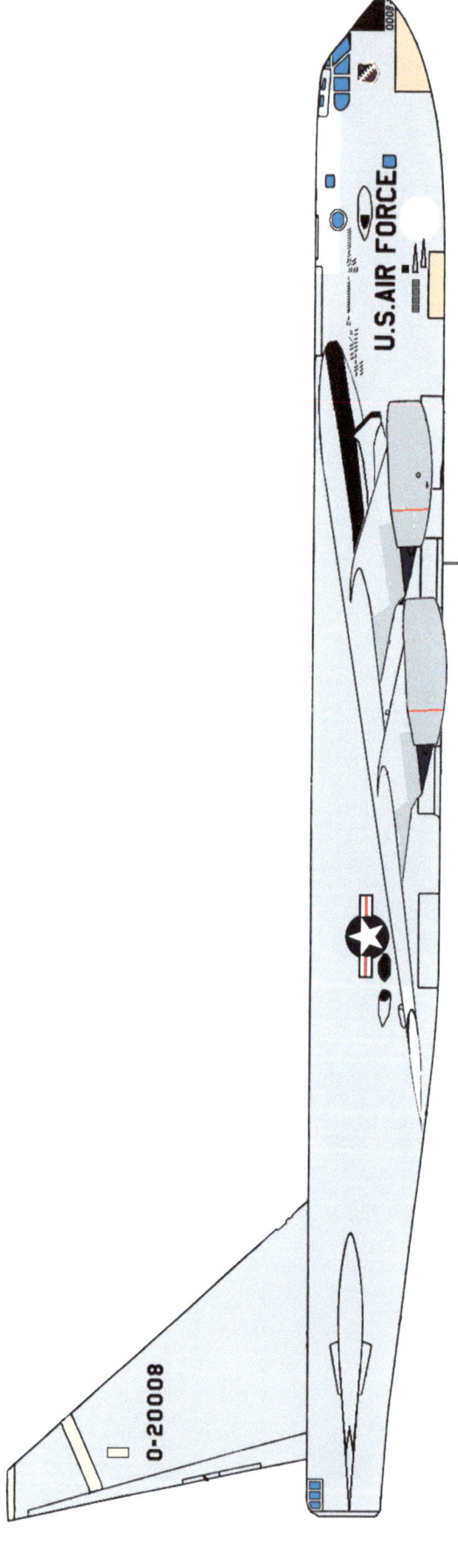

@ 1971

Sometime before May 1970, the X-15 mission marks were stripped and replaced with three marks for the last powered fligh and two captive carry missions. Three sets of marks were added for the X-24A, M2-F2, and HL-10 lifting bodies. Bruce Peterson's crash landing in the M2-F2 was represented by an inverted mark, and the gray dielectric panels on the vertical stabilizer were either painted white or replaced with white panels.
Sometime after May 1970, the nose art was temporarily removed from the right side of the fuselage leaving a circle with slightly different reflectance.

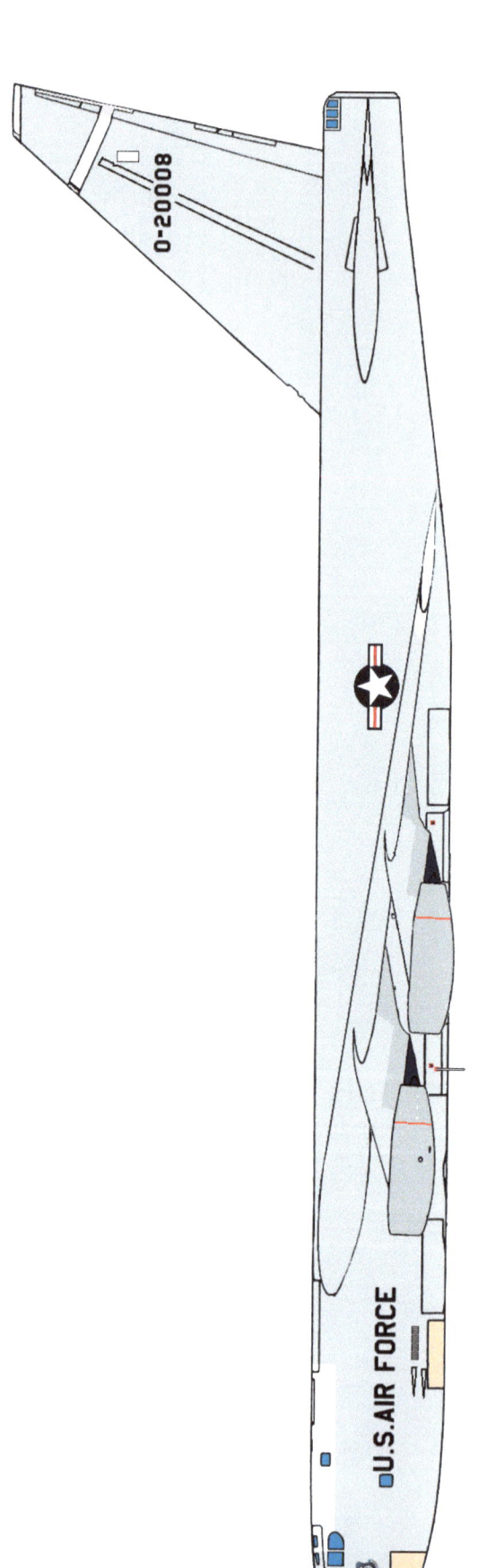

@ *May 1972*

The buff colored chin radome was replaced with a black radome.
DayGlo orange trim was added to the underside of the right wing tip.

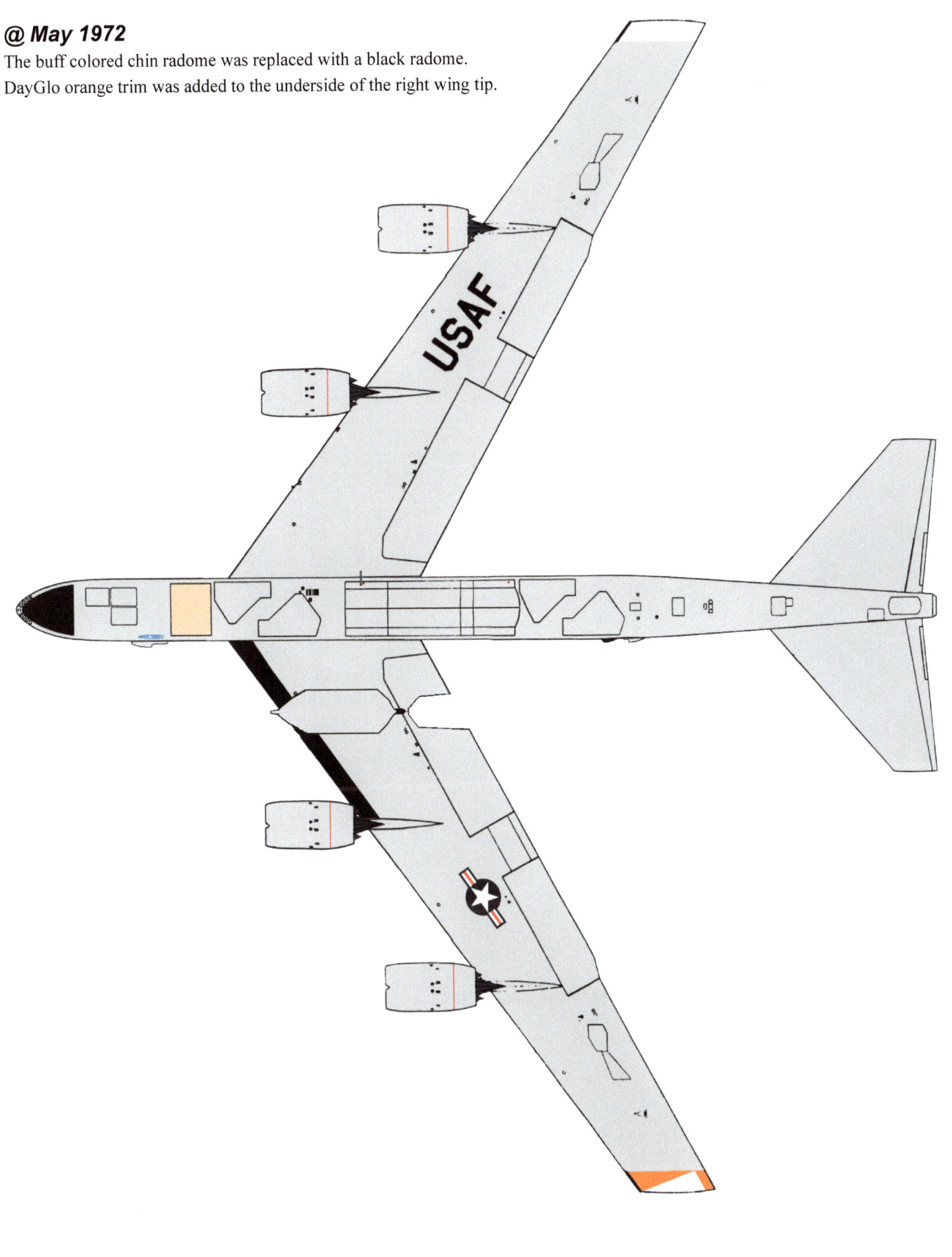

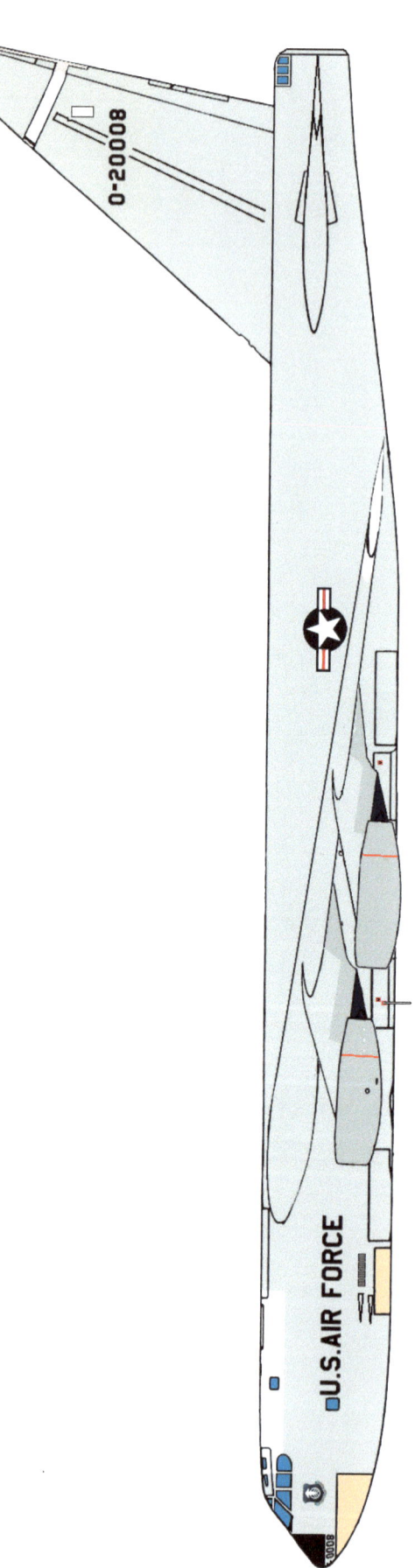

The nose art was restored, but the mission marks were removed.

@December 1974

The buff colored belly radome was replaced with a black radome.
The DayGlo orange trim on the underside of the right wing tip was removed.

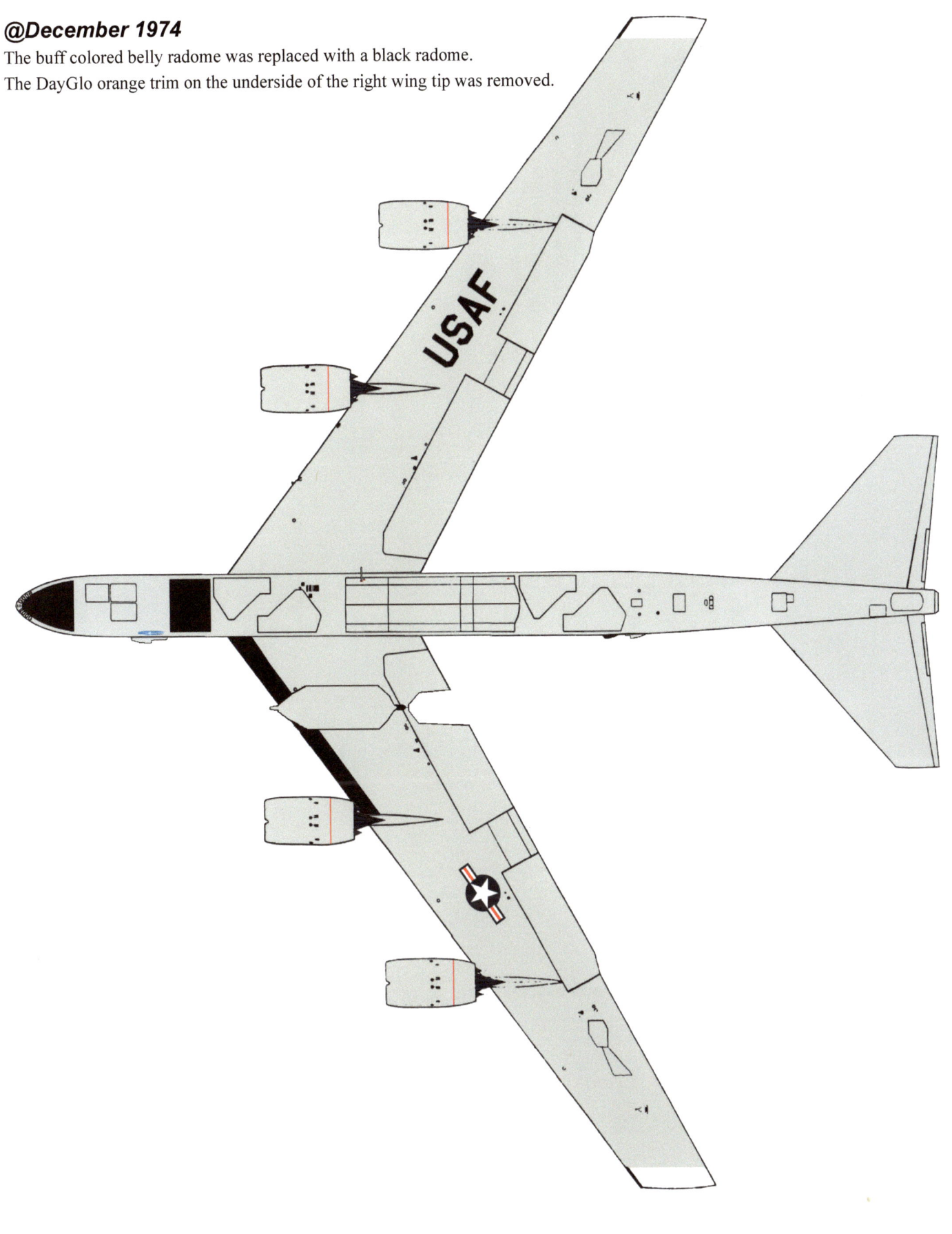

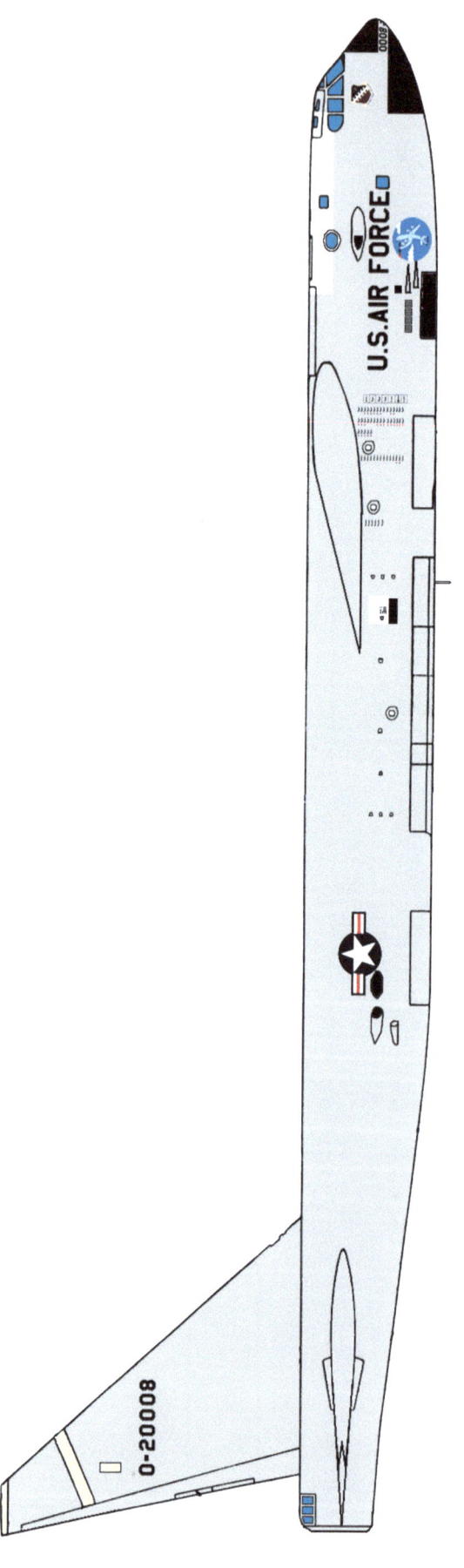

New mission marks were installed sometime before May 1973. The X-15, M2-F2, HL-10, and X-24A were each represented by a single mission mark and the number of flights launched from the NB-52B. M2-F3 and F-15 RPRV mission marks were added to the left of the mission tallies and mission marks for subsequent programs continued along the side of the fuselage.

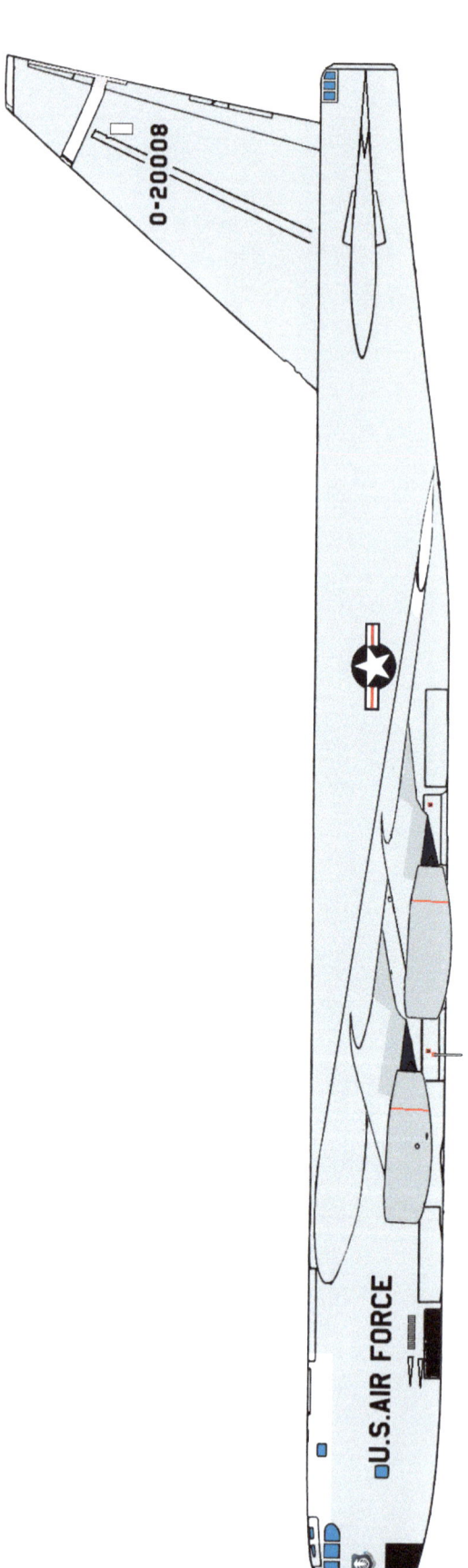

@1977

The NB-52B was transferred from the Air Force to NASA on April 26, 1976.
The flaps and rear portion of the underside of the wings were painted light gray.
The right wing inboard of the X-15 pylon was not painted.

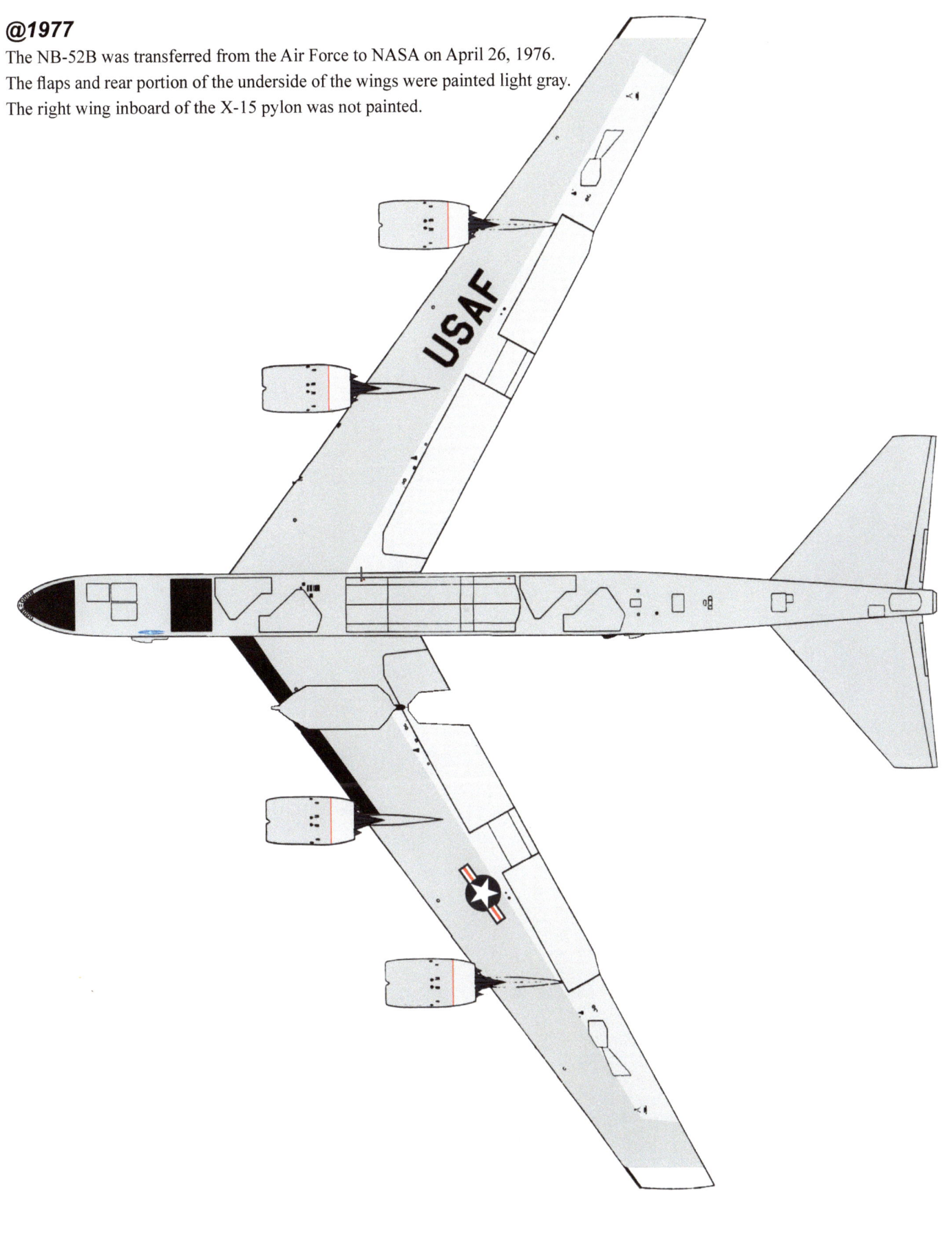

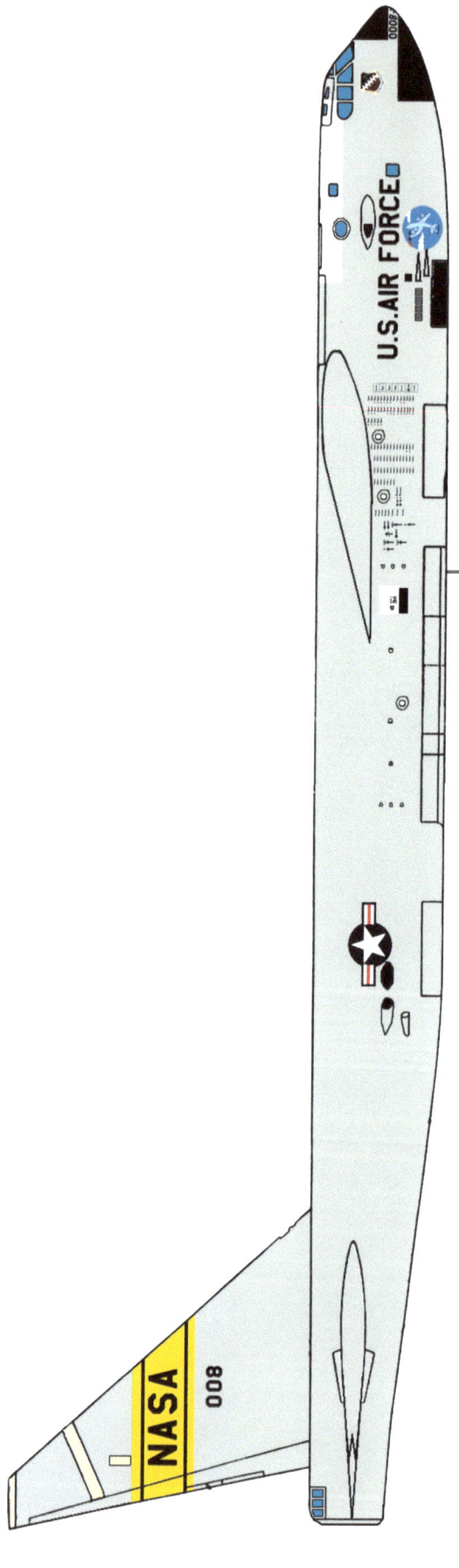
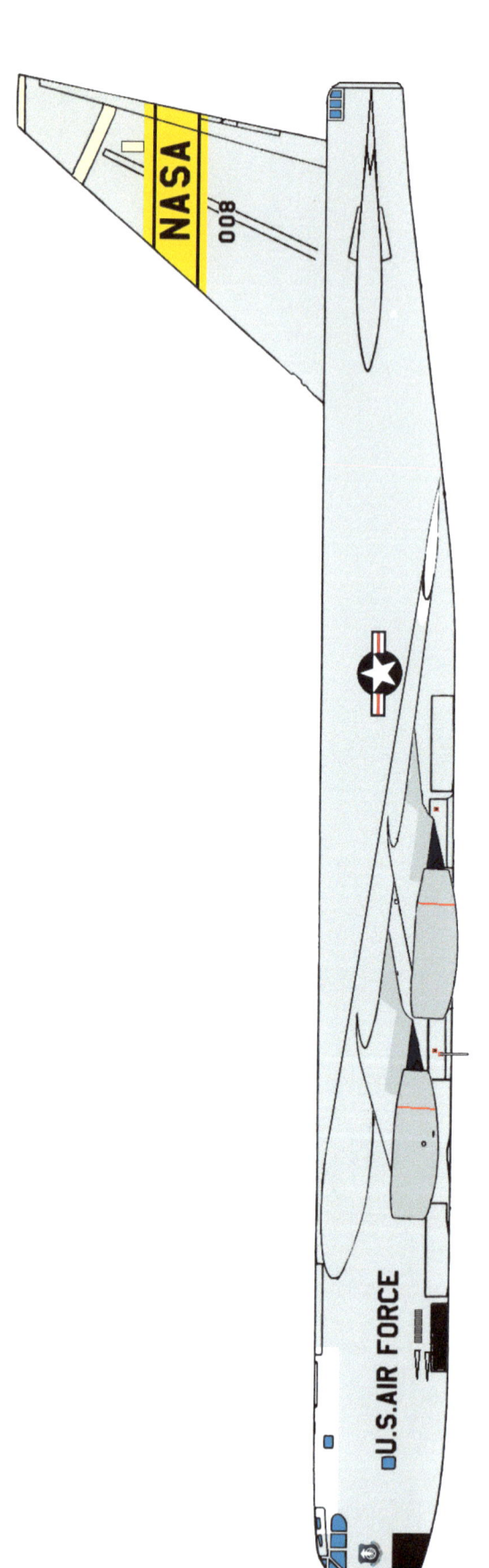

A six-foot white yellow band was added to the vertical stabilizer. The acronym NASA was placed in the center of the yellow band in thirty-six inch tall letters. The tail band was accented with six inch wide black stripes above and below the NASA acronym. The serial number on the tail was contracted from 0-20008 to 008.

@ 1979
The bomb bay liquid oxygen tanks were removed. The jettison pipe and overflow vents were removed from the left side of the bomb bay.

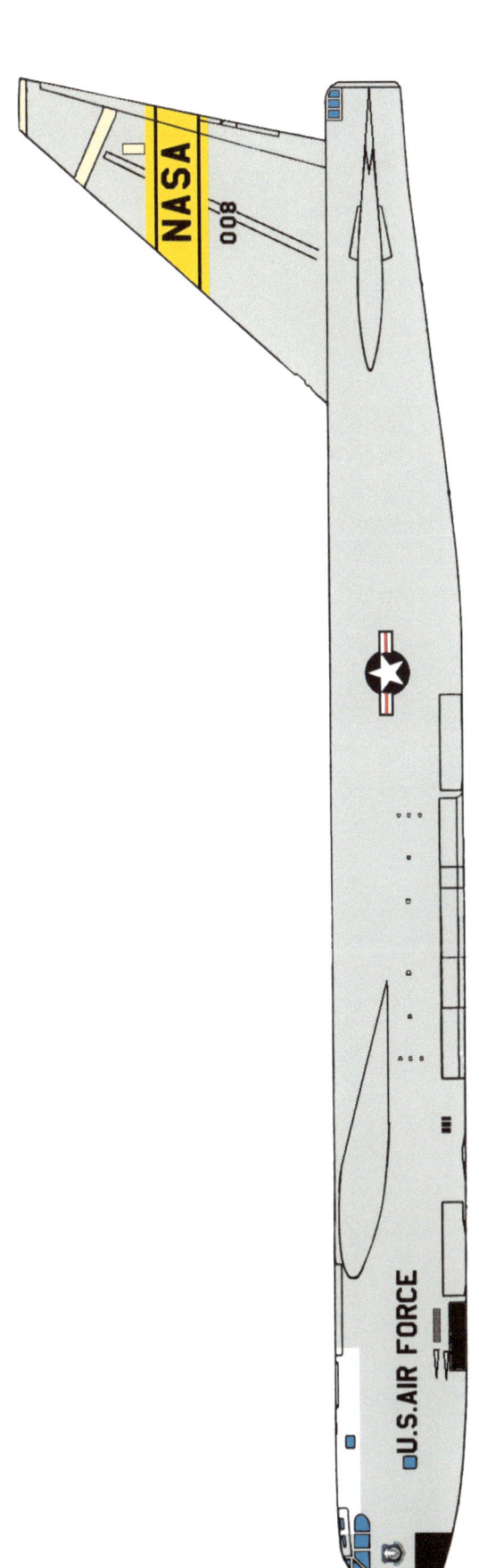

@ June 1995

The X-15 pylon was removed.

The notch in the trailing edge of the wing was fitted with an insert. The insert was painted zinc chromate green.

A test framework bearing a pair of J85 jet engines was temporarily installed in the bomb bay.

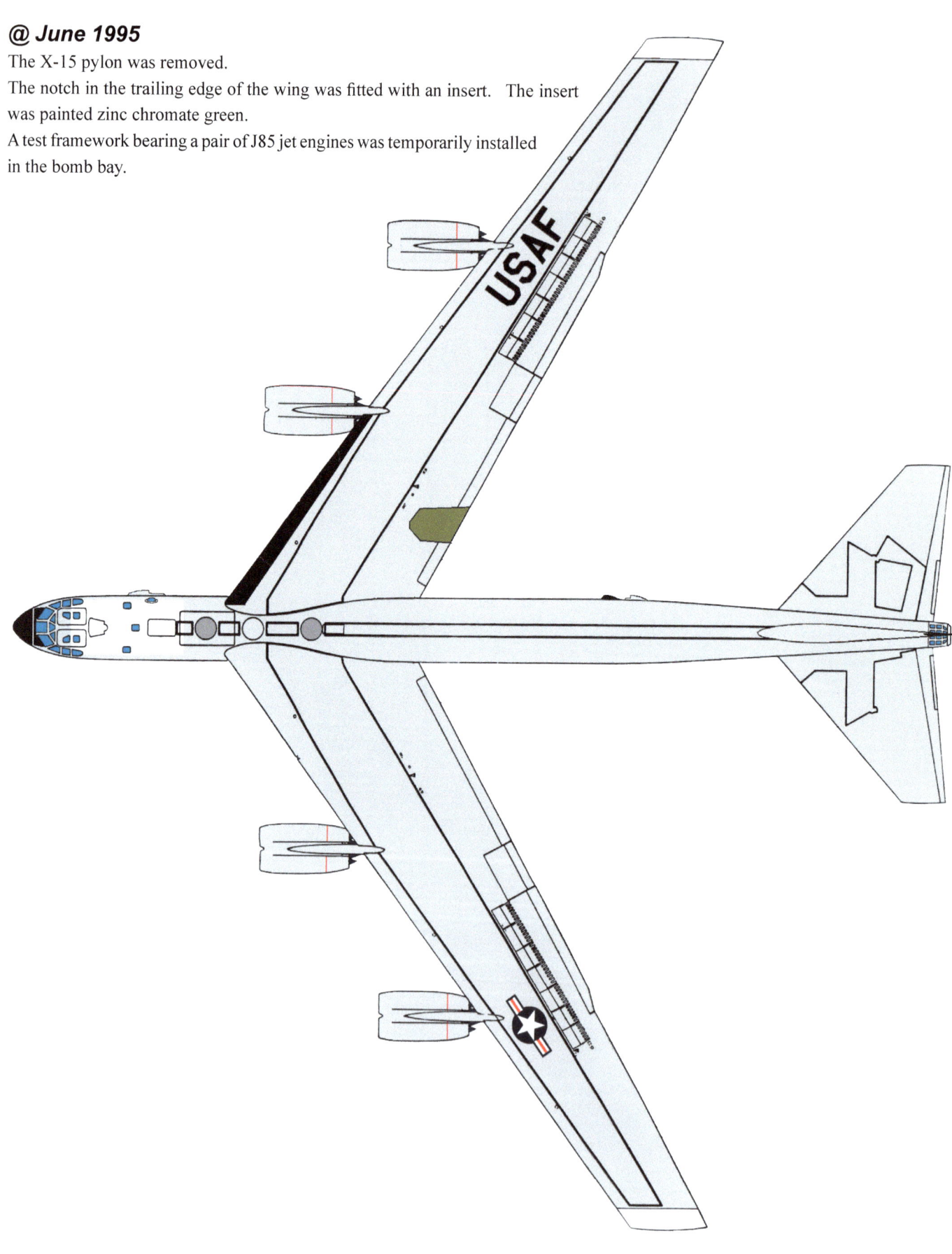

The gray painr on the trailing edge of the underside of the wing had been removed or faded away since 1977.

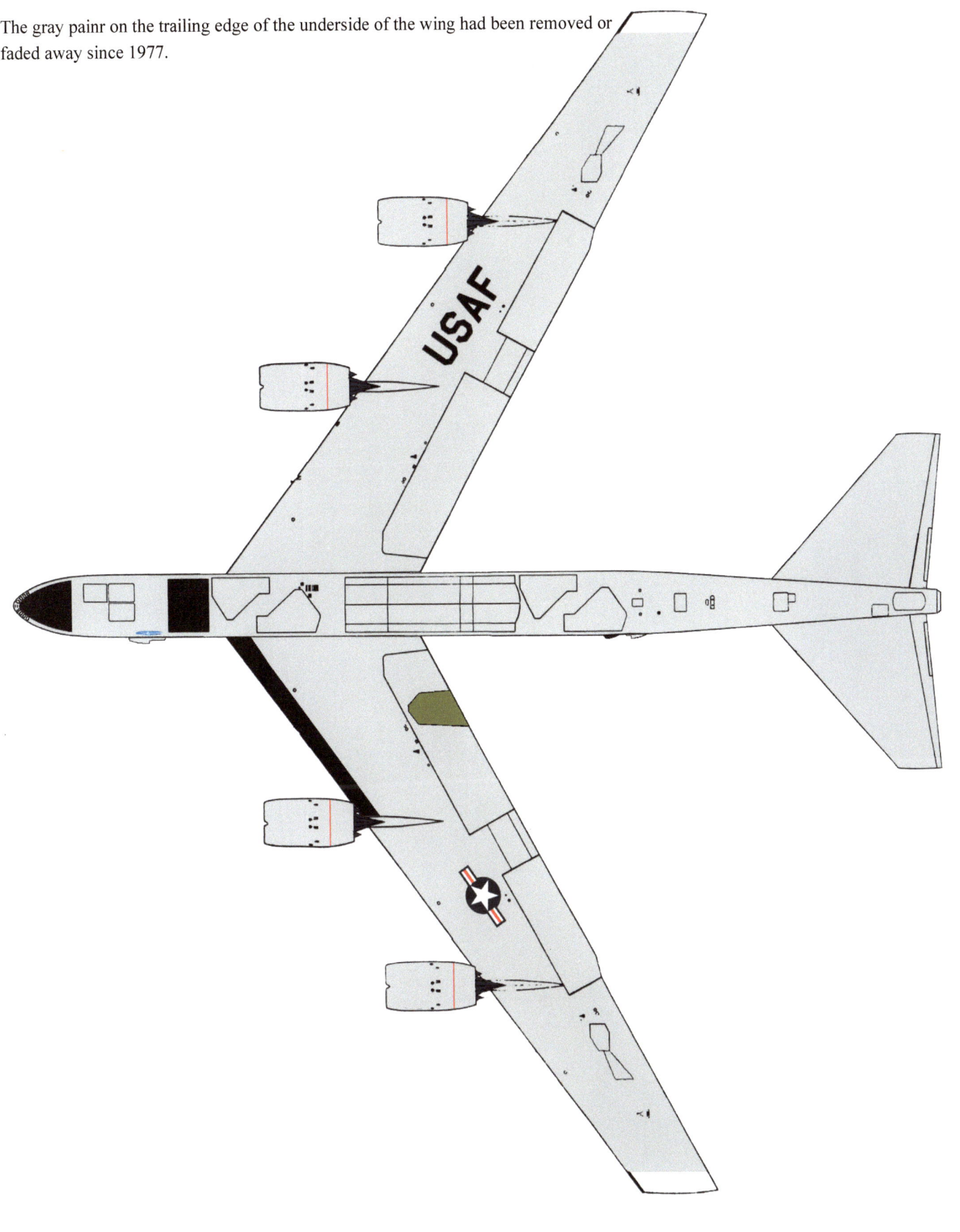

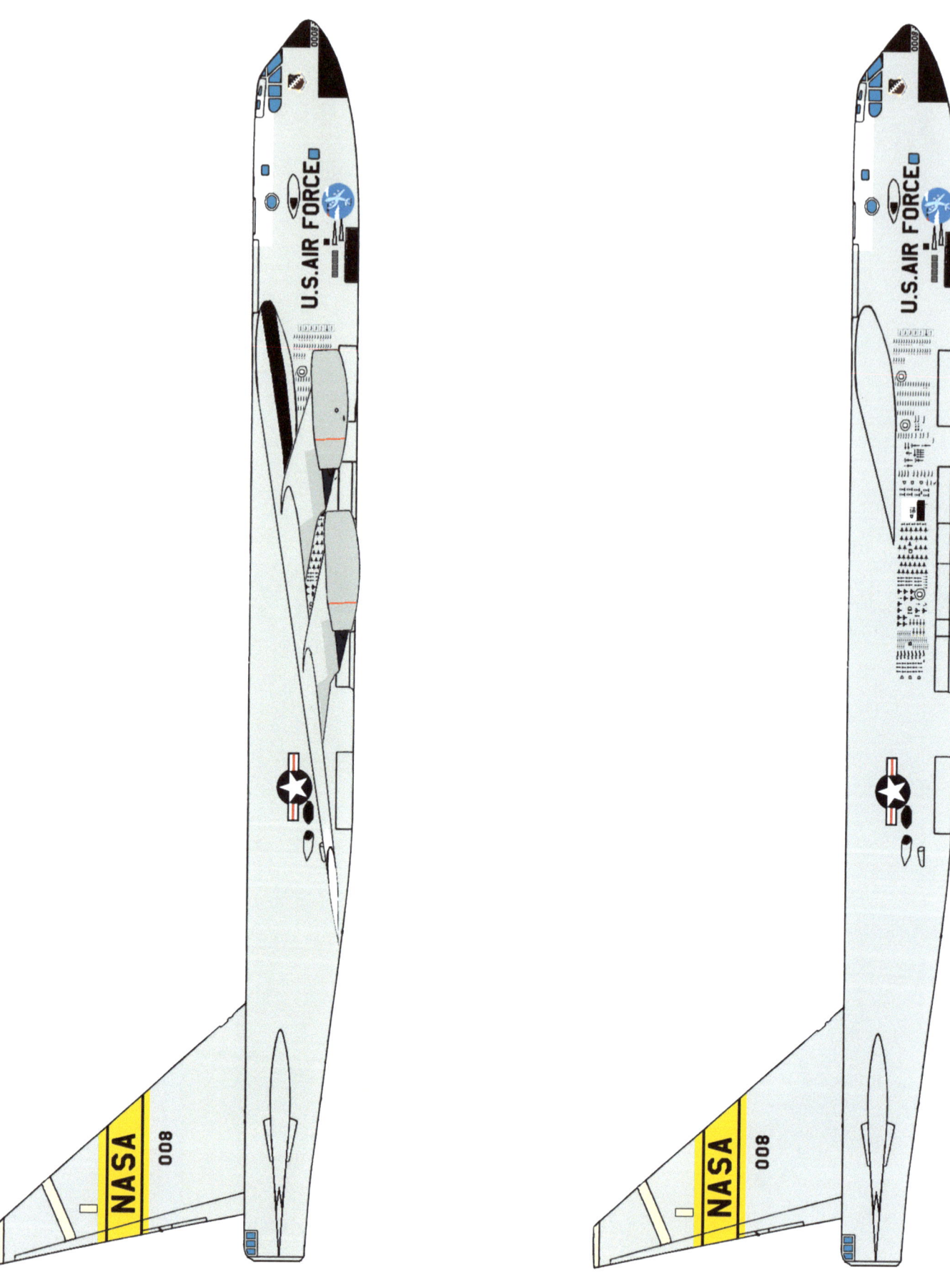

@ April 2001

The X-15 pylon was reinstalled for the X-43A Hyper-X program. The inboard side of the pylon was painted black and the outboard side was painted white. The insert in the wing slot was reduced in size to provide clearance for the tail of the Hyper-X stack.

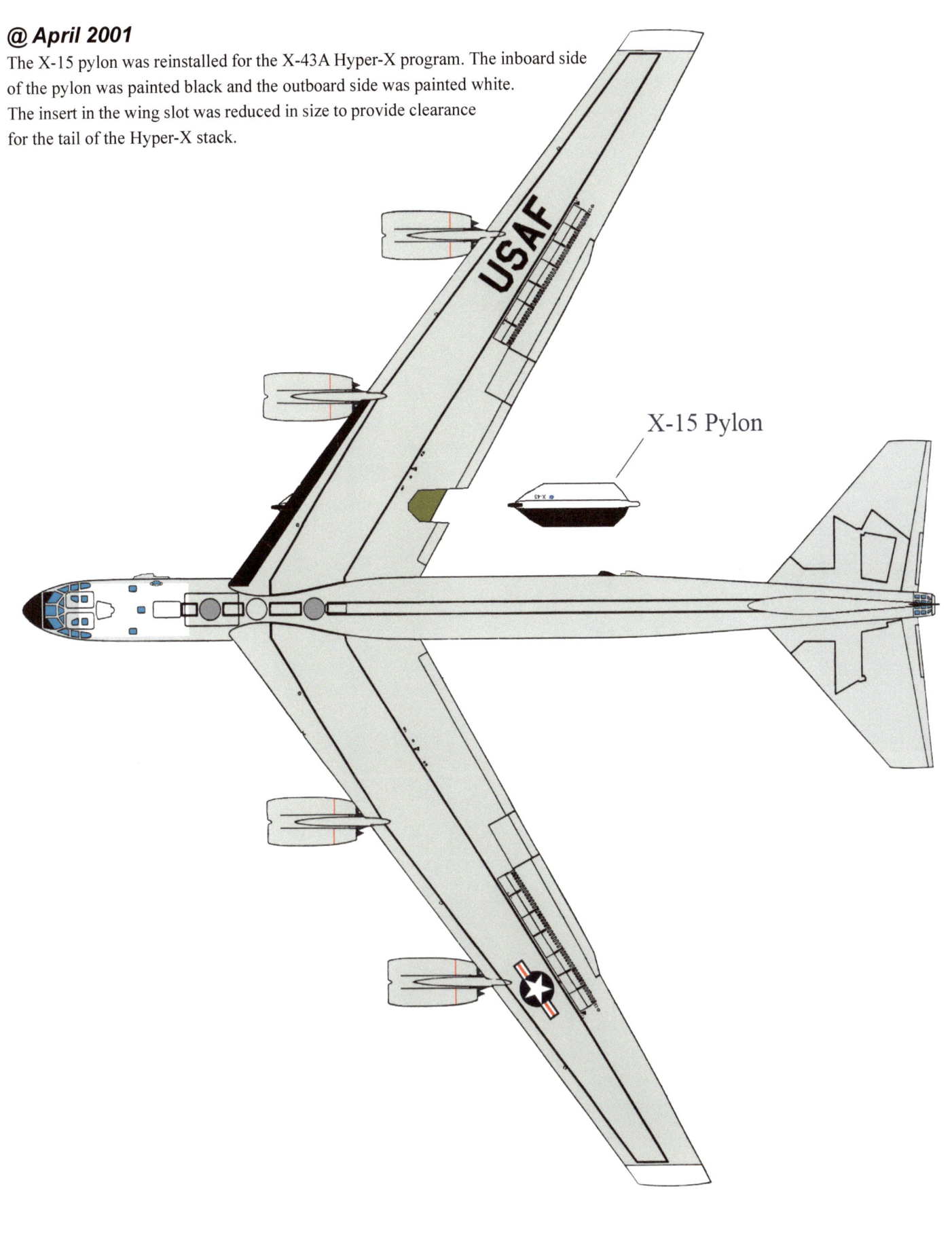

X-15 Pylon

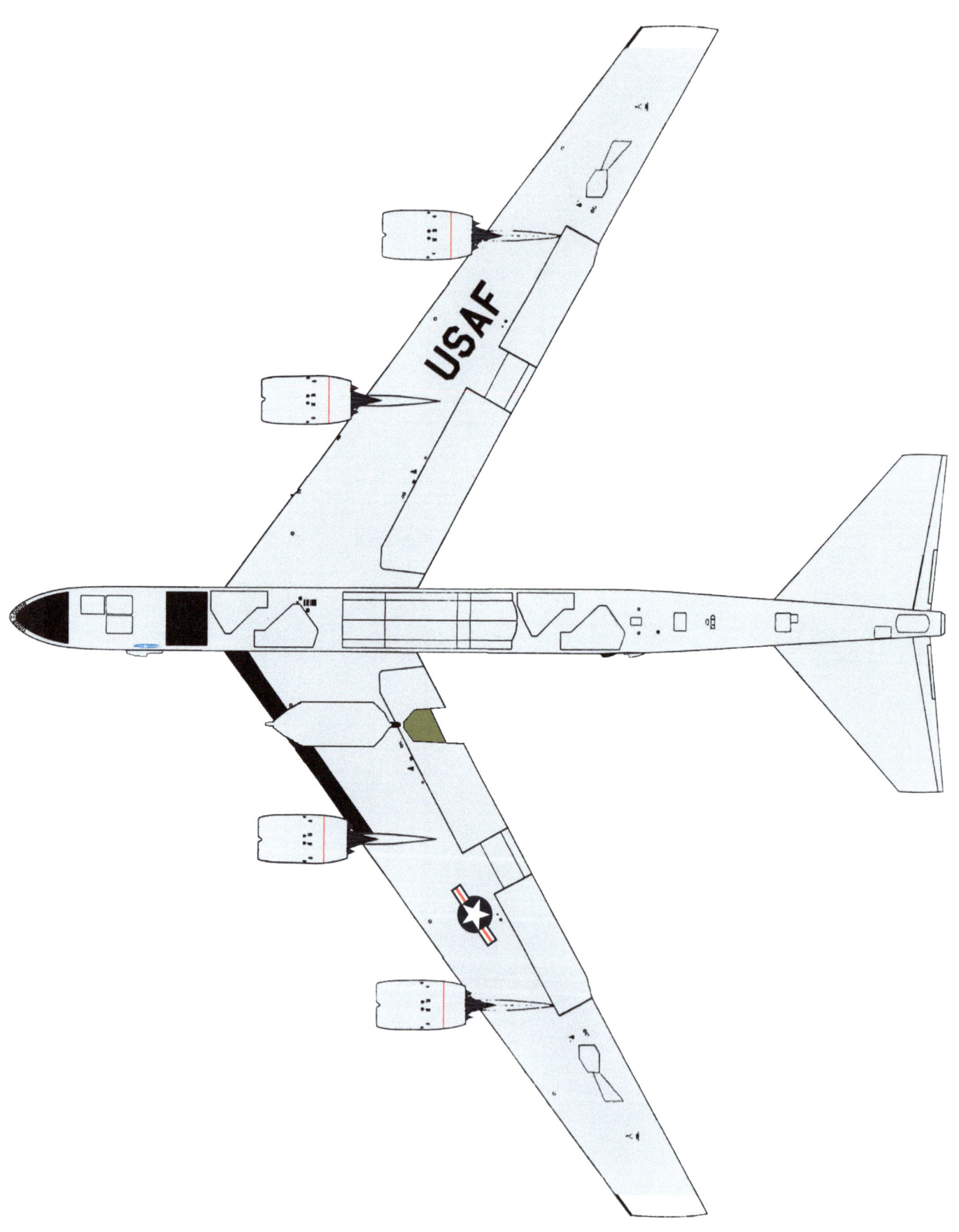

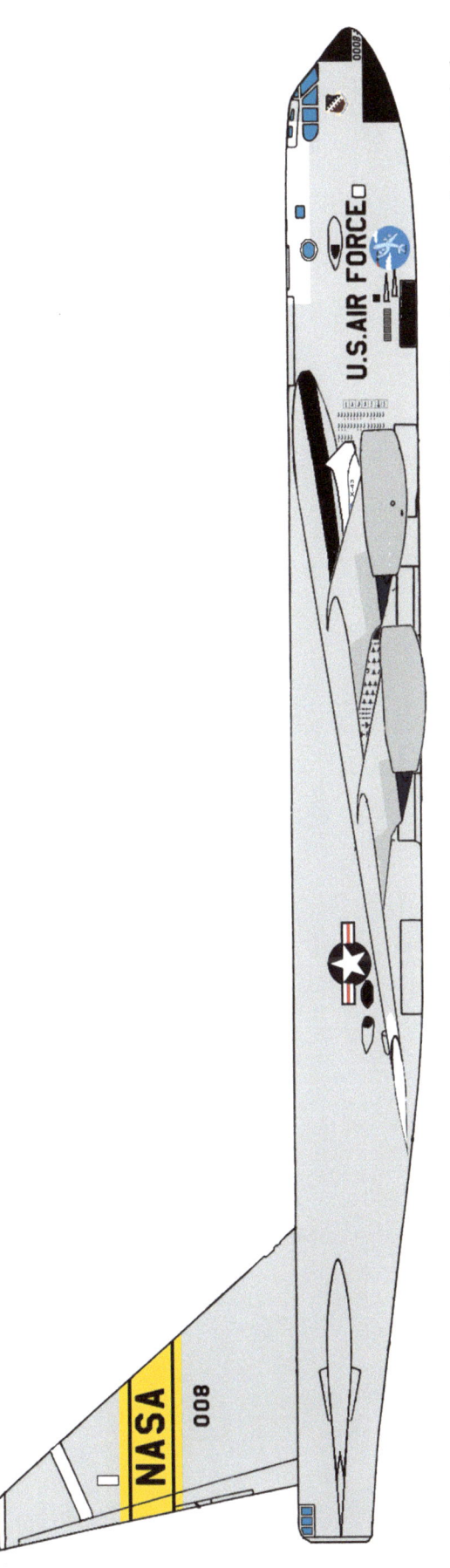

The windows on the lower deck were replaced with opaque panels.

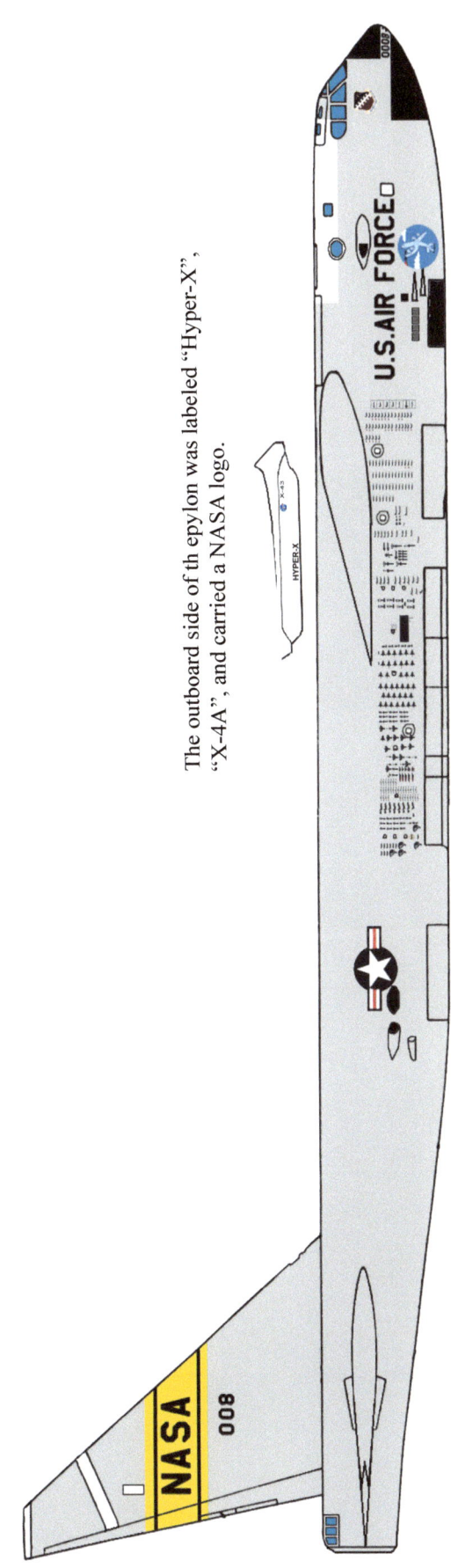

The outboard side of th epylon was labeled "Hyper-X", "X-4A", and carried a NASA logo.

@ December 17, 2004

The NB-52B made its last flight on November 16, 2004..
It was retired in a ceremony at the NASA Dryden Flight Research Center on December 17, 2004.

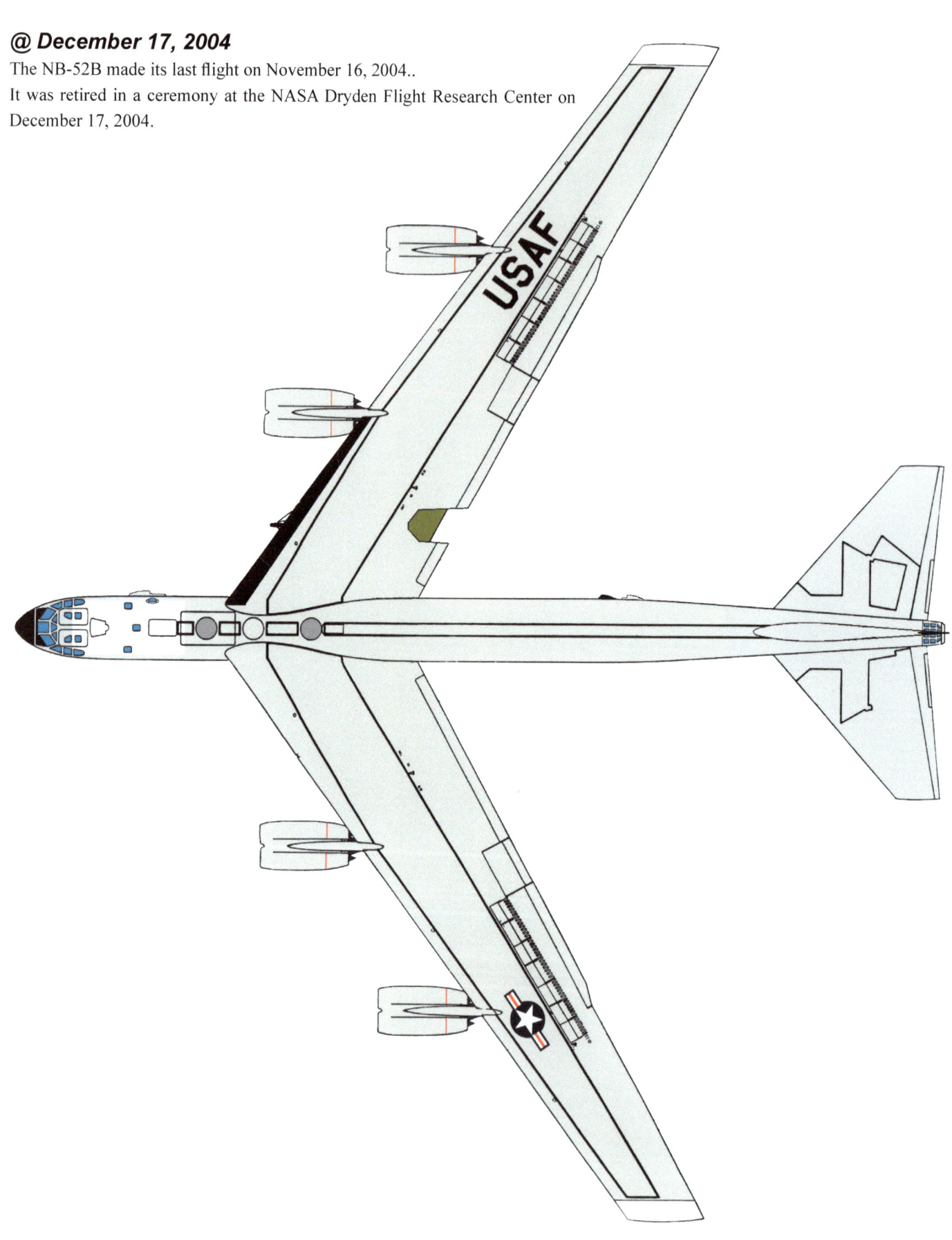

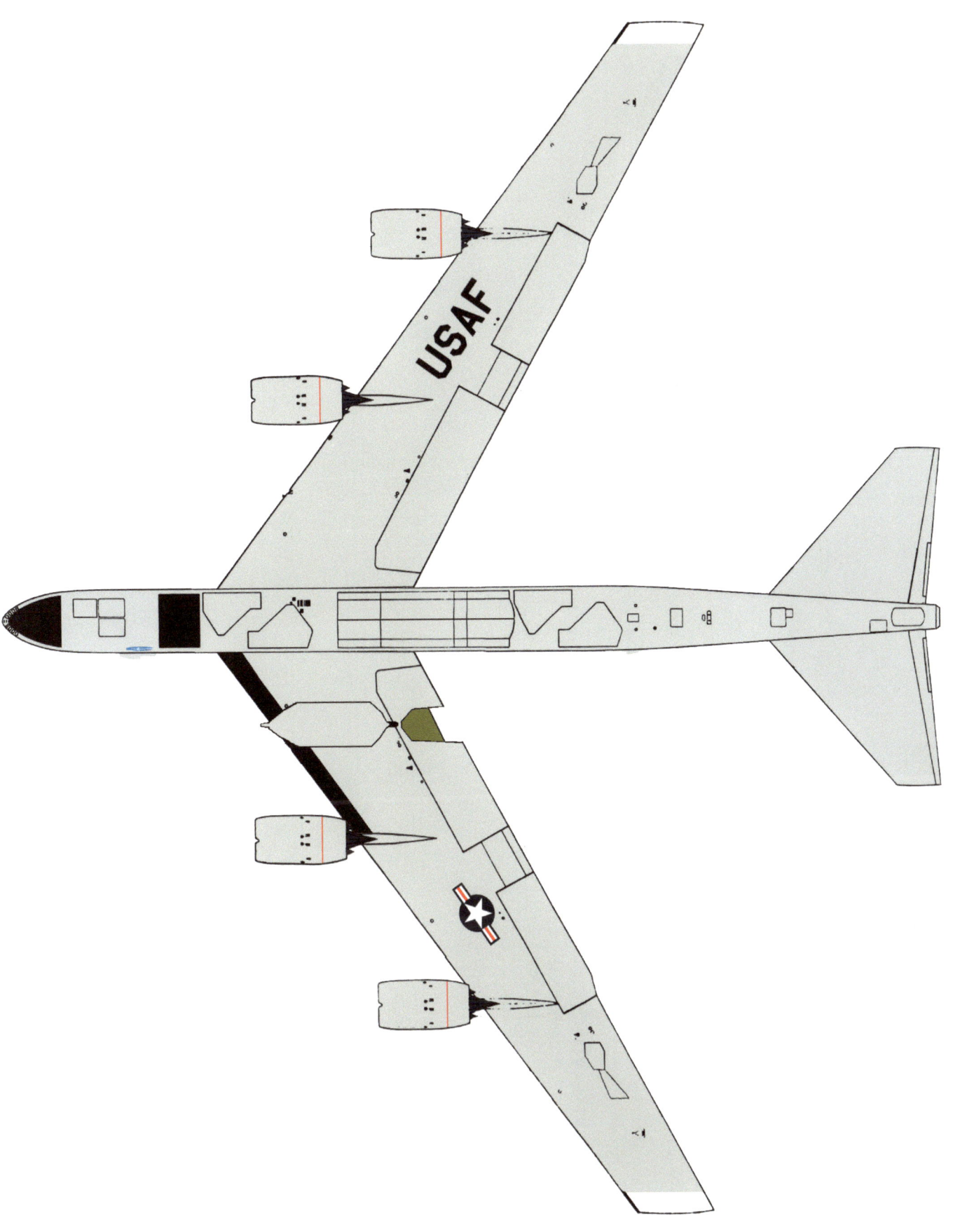

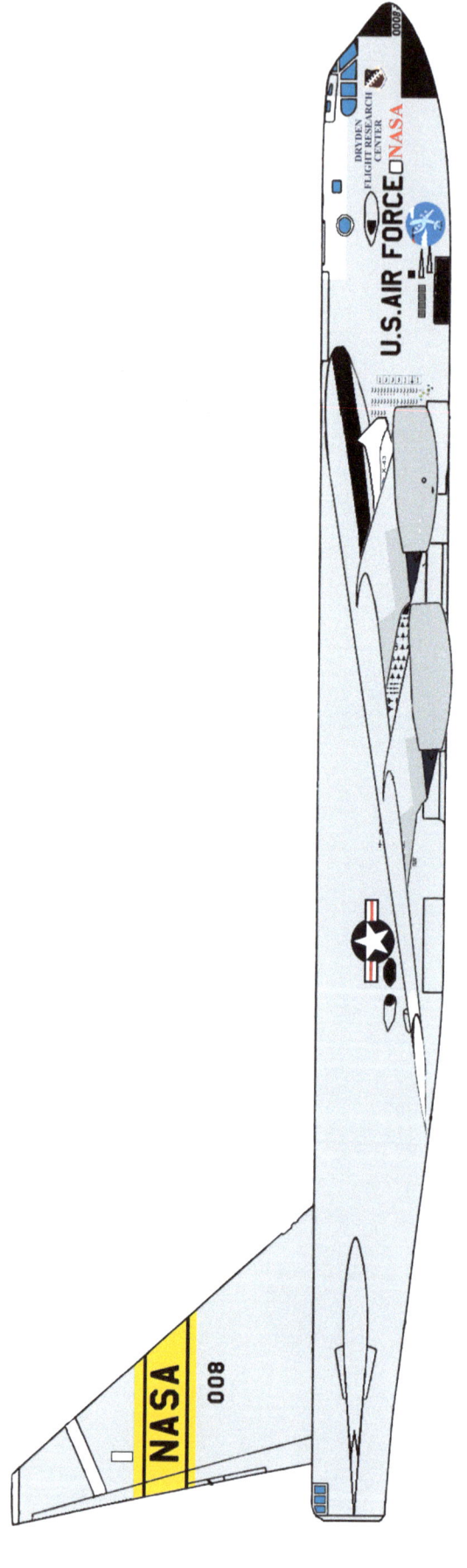

"DRYDEN FLIGHT RESEARCH CENTER" was added to the nose in blak lettering for the last flight. "NASA" was added to the nose in red lettering. A collection of decals had been applied below the M2-F3 mission marks. "T-THA...T-THAT'S ALL FOLKS!" was added to the left of the mission marks following the final flight.

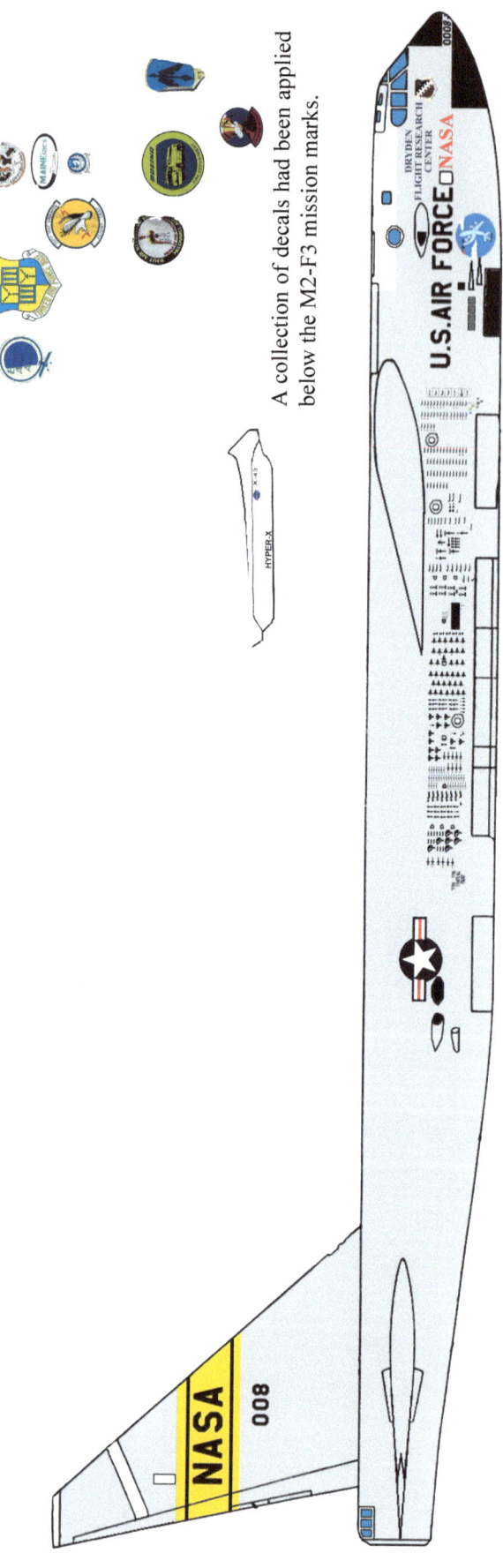

A collection of decals had been applied below the M2-F3 mission marks.

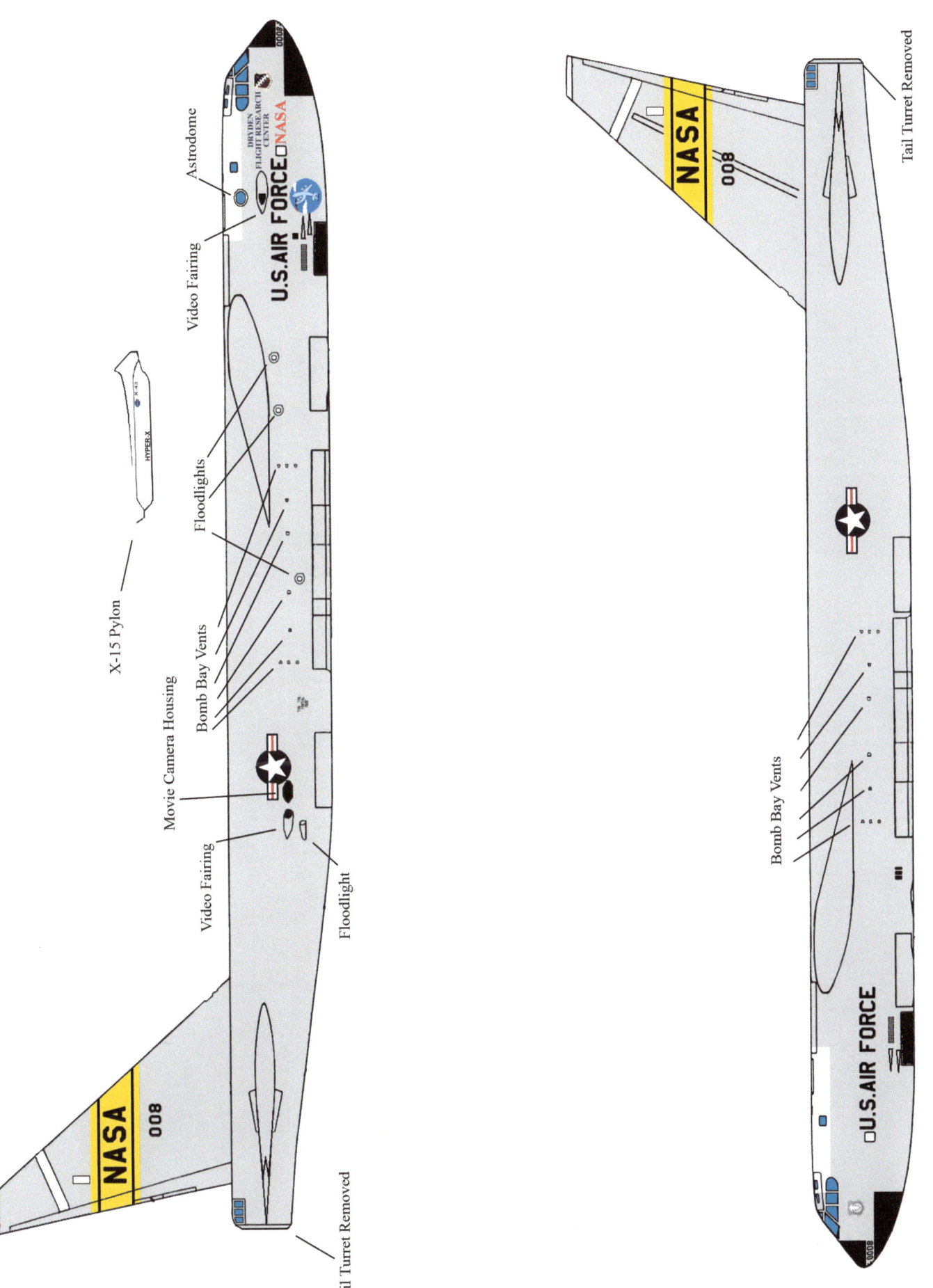

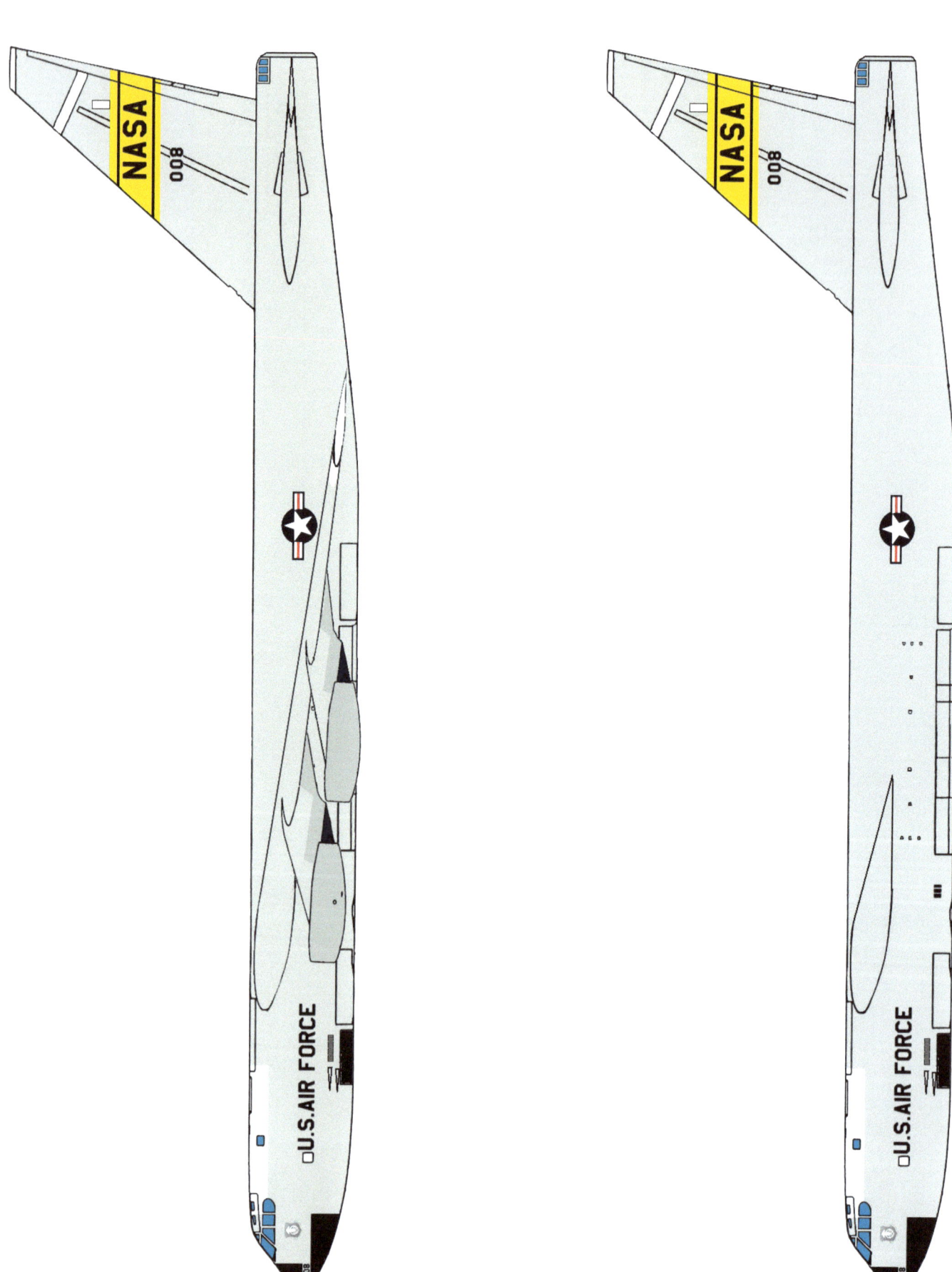

www.ingramcontent.com/pod-product-compliance
Lightning Source LLC
Chambersburg PA
CBHW040912020526
44116CB00026B/34